AF845345

Practical Application of Online Monitoring

Awwa
Research
Foundation

About the Awwa Research Foundation

The Awwa Research Foundation (AwwaRF) is a member-supported, international, nonprofit organization that sponsors research to enable water utilities, public health agencies, and other professionals to provide safe and affordable drinking water to consumers.

The Foundation's mission is to advance the science of water to improve the quality of life. To achieve this mission, the Foundation sponsors studies on all aspects of drinking water, including supply and resources, treatment, monitoring and analysis, distribution, management, and health effects. Funding for research is provided primarily by subscription payments from approximately 1,000 utilities, consulting firms, and manufacturers in North America and abroad. Additional funding comes from collaborative partnerships with other national and international organizations, allowing for resources to be leveraged, expertise to be shared, and broad-based knowledge to be developed and disseminated. Government funding serves as a third source of research dollars.

From its headquarters in Denver, Colorado, the Foundation's staff directs and supports the efforts of more than 800 volunteers who serve on the board of trustees and various committees. These volunteers represent many facets of the water industry, and contribute their expertise to select and monitor research studies that benefit the entire drinking water community.

The results of research are disseminated through a number of channels, including reports, the Web site, conferences, and periodicals.

For subscribers, the Foundation serves as a cooperative program in which water suppliers unite to pool their resources. By applying Foundation research findings, these water suppliers can save substantial costs and stay on the leading edge of drinking water science and technology. Since its inception, AwwaRF has supplied the water community with more than $300 million in applied research.

More information about the Foundation and how to become a subscriber is available on the Web at **www.awwarf.org**.

Practical Application of Online Monitoring

Prepared by:
Michelle M. Frey and **Laurie Sullivan**
McGuire Environmental Consultants, Inc.
1855 Blake Street, Denver, CO 80202

Sponsored by:
Awwa Research Foundation
6666 West Quincy Avenue, Denver, CO 80235-3098

Published by:

Awwa Research Foundation
Advancing the Science of Water®

IWA Publishing

DISCLAIMER

This study was funded by the Awwa Research Foundation (AwwaRF). AwwaRF assumes no responsibility for the content of the research study reported in this publication or for the opinions or statements of fact expressed in the report. The mention of trade names for commercial products does not represent or imply the approval or endorsement of AwwaRF. This report is presented solely for informational purposes.

Published by IWA Publishing, Alliance House, 12 Caxton Street, London SW1H 0QS, UK

Telephone: +44 (0) 20 7654 5500; Fax: +44 (0) 20 7654 5555; Email: publications@iwap.co.uk;
Web: **www.iwapublishing.com**

AwwaRF report number 91022F. Originally published by AwwaRF for its subscribers in 2004.
IWA Publishing version published 2005.

ISBN 1 84339 912 1

CONTENTS

TABLES

FIGURES

FOREWORD

The Awwa Research Foundation is a nonprofit corporation that is dedicated to the implementation of a research effort to help utilities respond to regulatory requirements and traditional high-priority concerns of the industry. The research agenda is developed through a process of consultation with subscribers and drinking water professionals. Under the umbrella of a Strategic Research Plan, the Research Advisory Council prioritizes the suggested projects based upon current and future needs, applicability, and past work; the recommendations are forwarded to the Board of Trustees for final selection. The foundation also sponsors research projects through an unsolicited proposal process; the Collaborative Research, Research Applications, and Tailored Collaboration programs; and various joint research efforts with organizations such as the U.S. Environmental Protection Agency, the U.S. Bureau of Reclamation, and the Association of California Water Agencies.

This publication is a result of one of these sponsored studies, and it is hoped that its findings will be applied in communities throughout the world. The following report serves not only as a means of communicating the results of the water industry's centralized research program but also as a tool to enlist the further support of the nonmember utilities and individuals.

Projects are managed closely from their inception to the final report by the foundation's staff and large cadre of volunteers who willingly contribute their time and expertise. The foundation serves a planning and management function and awards contracts to other institutions such as water utilities, universities, and engineering firms. The funding for this research effort comes primarily from the Subscription Program, through which water utilities subscribe to the research program and make an annual payment proportionate to the volume of water they deliver and consultants and manufacturers subscribe based on their annual billings. The program offers a cost-effective and fair method for funding research in the public interest.

A broad spectrum of water supply issues is addressed by the foundation's research agenda: resources, treatment and operations, distribution and storage, water quality and analysis, toxicology, economics, and management. The ultimate purpose of the coordinated effort is to assist water suppliers to provide the highest possible quality of water economically and reliably. The true benefits are realized when the results are implemented at the utility level. The foundation's trustees are pleased to offer this publication as a contribution toward that end.

The use of online monitoring devices by water utilities around the world is gaining widespread acceptance and, indeed, is often required by regulatory programs. Developing a best practices program to support the application of these devices is necessary for drinking water utilities. The lack of existing standards and certification programs to direct and guide utilities on online instrument selection, maintenance, calibration, and use can lead to confusion. This study provides practical guidelines for utilities in the application of online instruments and the monitoring data collected by these instruments.

Walter J. Bishop
Chair, Board of Trustees
Awwa Research Foundation

James F. Manwaring, P.E.
Executive Director
Awwa Research Foundation

FOREWORD

ACKNOWLEDGMENTS

The authors of this report would like to acknowledge the contributions made by the Project Advisory Committee members:

- Wayne Geriak, New York City Department of Environmental Protection
- Kerwin Rakness, Process Applications Inc., Fort Collins, Colo.
- John Wierenga, City of Grand Rapids Water System, Michigan

Special thanks are due to Emerson Lomaquahu and Ryan Ulrich, Awwa Research Foundation project managers. In addition, we appreciate the participation and contributions made by our case study utilities.

We also wish to acknowledge the assistance of Shana DeBault of Stratus Consulting, Inc., Boulder, Colo., in the preparation of this manuscript.

EXECUTIVE SUMMARY

For this project, the status of online instrument use in U.S. water utilities was surveyed. The results indicated that particle analyzers and oxidant analyzers are the most commonly utilized online instruments. More uncommon analyzers, such as phosphate, gas chromatograph, and ultraviolet light were identified by just a few utilities. These may be an indication of the upcoming trends in online instrument use.

Case studies of six utilities were performed to supplement the survey results by recording, in-depth, the instrument uses, maintenance practices, instrument support structures, and supervisory structures of the six utilities as related to online instrument use. In addition, the case study utility supplied online and bench data from their online instruments for analysis. The utilities selected included conventional filtration, direct filtration, lime softening, and unfiltered systems with widely varying source water quality. The case study results indicate a variety of different practices and uses surrounding online instruments; however, a majority of the instruments are used for trending and informational purposes. In the comparison of bench versus online instruments, systematic bias in the measurements was observed. For example, in the case of online turbidimeters, online measures were consistently lower than the bench measures.

The application of data generated from online instruments is discussed in detail, and practical methods for improving or enhancing data use are presented. A template for developing standard operating procedures (SOPs) was created during this project. The SOP template includes specific instructions on how to determine a calibration frequency based on instrument application, location, and desired accuracy. Both the template and the approach are comprehensive and offer utilities a straightforward approach to SOP development.

CHAPTER 1
INTRODUCTION

Even though many utilities use online monitoring to control and monitor treatment processes—and to serve as the basis for important management decisions—little information is available to guide utilities through the installation, operation, maintenance, and overall application of online sensors. Critical parameters such as flow, turbidity, pH, and chlorine residual are commonly monitored using online instrumentation. The data reported from these instruments are essential to plant operations; for example, the information may be used to pace feed rates for coagulants or disinfectants or to calculate disinfectant contact time. The data may also be submitted on compliance reports to regulating agencies.

This study examined firsthand how utilities are currently utilizing and supporting online instruments. In addition, guidelines were developed for utilities to consider as part of their online monitoring programs. For U.S. utilities, in particular, this study may support online monitoring applications for recently promulgated regulatory requirements. Other uses of online monitoring data are increasing at water utilities, as the utilities increasingly demand high-quality data collected during short intervals to support such activities as treatment plant automation, operator notification, treatment need forecasting, and early warning of potential water quality challenges.

Advances in online instrumentation will continue. Data from additional water quality parameters will become increasingly available through online instruments, and the supervisory control and data acquisition (SCADA) systems to which they are linked will become more sophisticated. These trends will impart significant changes in treatment plant operation practices. Specifically, many of the functions currently performed by plant staff will become amenable to computer control. Operations staff priorities are likely to shift accordingly, and the overall support of online instruments will become increasingly critical to successful utility operations. As the future of water treatment inevitably moves toward systems that rely heavily on online monitoring, establishing application, installation, operation, and maintenance guidelines becomes prudent.

BACKGROUND ON ONLINE SENSORS

In today's water utility, online instruments are important process control and reporting tools. The regulations being published, such as the Interim Enhanced Surface Water Treatment Rule, only increase the emphasis that will be placed on the importance of online instruments. Despite these trends, a literature review on the operation and application of online instruments revealed very little information on these topics. Published information primarily focuses on the application of particle analyzers, namely turbidimeters and particle counters. The fact that particle analyzers are the most discussed instrument class in the literature is not surprising because this is one of the most prevalent instrument types at water utilities, as reported in chapter 2 of this study.

In a study performed by Burlingame, Pickel, and Roman (1998), the authors raised important concerns about application of turbidity, specifically low level, for performance trending versus compliance monitoring (maximum contaminant level [MCL]). Based on the currently available quality assurance–quality control (QA–QC) requirements, using turbidity as a treatment technique requirement has been suitable. However, if turbidity was used as an MCL, the current available technology might not be adequate, both because low-level readings are not yet reliable and because standardized methods and the associated proper QC measures have not been developed for the level that an MCL would require. The authors warn that taking turbidity to a level at which accuracy is important without supporting it with the appropriate QA–QC measures would be "inappropriate."

In addition, the Burlingame, Pickel, and Roman (1998) report specifically raises difficulties surrounding the application of online turbidimeters, including the fact that different models have different data collection rates, sample chamber volumes, and sample flow rates. Variations in these factors can result in differences in the duration and magnitude of turbidity peaks and in how they are displayed graphically.

Particle counting has been shown to be a very valuable tool for evaluating the effects of process changes. In a study performed by Lewis, Hargesheimer, and Yentsch (1992), particle counting proved to be a good indication of filter and plant performance. The same study indicated that turbidity values in the range of 0.03 ntu to 0.10 ntu did not respond to small changes in coagulant dosage. Particle counting, then, proved to be a better tool for monitoring filtered and finished water quality. However, it should be noted that the ability to compare performance between particle counters has been under scrutiny for some time (VanGelder, Chowdhury, and Lawler 1999). Based on these limitations, regulatory compliance applications would not be suitable. In addition, direct comparison of, for example, filter effluent from two different particle counters may also be suspect. However, trending of an individual filter using the same particle counter could yield very important information about performance. Also, Lewis, Hargesheimer, and Yentsch (1992) raised the concern that important design aspects of particle counters can vary. These aspects, including operation principles, resolution, data processing power, and physical installation requirements, should all be considered before selecting a particle counter.

A number of researchers mentioned the importance of data management. The amount of data generated by online instruments is far greater than that generated by either a bench- or a batch-type instrument. Guidelines for data management were not specifically defined in the literature reviewed, including operations references and manufacturers information.

Another important deficiency identified in the search for information about online instruments was the lack of information presented in the water industry's "standard" operations reference materials. Although many larger utilities have instrument technicians or other personnel dedicated to calibrating and maintaining online instruments, the reality is that at many utilities, the operations staff is responsible for online instrument operation. A review of the resources to which operators typically refer, however, suggests otherwise. One manual stated that online instrumentation should be maintained by an "...instrumentation specialist or the factory representative rather than by an operator (unless specially instructed)." Operators have and will continue to train themselves on the instruments, using a combination of internal resources such as more experienced operators and the manufacturers' manuals.

Manufacturers' manuals are good documents for learning about an instrument, and a number of manuals were reviewed for this project. Each one contained comprehensive information on installation, calibration, and maintenance. However, the manuals have their limitations, most notably that they cannot deal with the full spectrum of conditions that will arise at all water treatment facilities. These must be identified and handled by plant personnel. For example, one manual clearly stated that a specific maintenance schedule could not be dictated because operating conditions can vary considerably, but that experience should make the intervals apparent. Plant-specific issues should be addressed by developing standard operating procedures (SOPs).

STANDARDS AND CERTIFICATION PROGRAMS

Standards for online instruments are provided only in terms of analytical method standardization procedures. Online instruments used for compliance monitoring in the United States are required to use analytical methods that the U.S. Environmental Protection Agency has approved for the regulated parameter. However, these instruments and their measurement performance are not subject to the same standard of laboratory certification required for bench-top analyses. Indeed, neither the American Society for Testing and Materials nor the International Standards Organization (ISO) has adopted any type of uniform standard or certification program for online instruments. The ISO has formed a technical committee to address this issue and a draft standard is under review by committee members.

In addition to standards for online instruments, using these instruments in the field requires specialized training and knowledge. As will be discussed in later chapters, the very largest utilities in the United States generally recognize this requirement and employ such technicians. However, the majority of water utilities that use online instruments use operations and maintenance staff to support these instruments. A review of operator certification programs for several states showed that there were few or no questions on online instrument maintenance, calibration, or application. The reality in most utility organizations is that training for online instrument support must be acquired outside of existing certification programs, and resources for such training is limited to those available through manufacturers or specialized contractors.

This lack of standardization and certification support for online instrument applications leads to confusion on the part of drinking water utilities about the best practices that they should implement when using online instruments. This study was specifically undertaken to assist water utilities in developing such best practices for their online monitoring programs.

APPLICATION ISSUES

Today, literally every industry uses online monitors. Many of the applications of online instruments in other industries are similar to the ways in which they are used in the water industry. Industries using water as a primary ingredient, such as the food, soft drink, and brewery industries, use the same monitoring equipment in much the same way as the water industry to meet high quality and reproducibility standards.

Many industries require purified water for their processes. The pharmaceutical and biotechnology industries use highly purified water to clean out process pipes before initiating product flow. Online monitors are used to monitor these streams to ensure overall process quality and cleanliness. Parameters measured include total hydrocarbons, pH, and particles. In the food industry, continuous sterilizers are used to produce clean, corrosion-free cans. The sterilizers are stainless steel and the sterilizing process results in a highly corrosive environment. To battle these conditions, online pH and chlorine residual are monitored and adjustments made to minimize the corrosion potential.

Although not widely used in the water industry, online detection devices for coliforms are currently available. Coliform measures are required at multiple locations in distribution systems on a monthly basis; for this reason, using online monitors for this purpose would be cost prohibitive at most utilities. However, if measuring coliforms as an indicator for other organisms in raw or finished water sources becomes practical, these monitors will become practical as well.

REPORT ORGANIZATION

The National Survey of Online Monitoring Practices at drinking water utilities is presented in chapter 2. The chapter also discusses the range of online instruments currently in use, as well as the support programs utilities have implemented to maintain their online monitoring activities. In chapter 3, findings from the various case study utility interviews and monitoring data are documented. Various methods for the evaluation of online monitoring performance are presented and used to evaluate several types of instruments in use at the case study utilities. Chapter 4 discusses the quality, use, and management of online monitoring data. Suggestions for real-time evaluation of online instrument performance are given along with strategies for the management of such information. Key to the success of a best practices program for online instruments is the development and implementation of a comprehensive SOP for each online instrument type in use at a utility. In chapter 5, the development of such SOPs is discussed, and the conclusions and recommendations generated by this study are presented.

CHAPTER 2
NATIONAL SURVEY OF ONLINE MONITORING PRACTICES

The National Survey of Online Monitoring Practices was undertaken to help understand current practices in online water quality monitoring by the U.S. drinking water industry. Specifically, this survey sought to answer the following questions:

- What types of analytical online instruments are being used?
- What are the predominant instrument applications?
- Are instruments regularly calibrated and are verification checks performed?
- At what frequency are checks performed and what are the trends among frequency, application, and instrument type?
- What are the trends among calibration frequencies and applications to instrument location?
- How many hours are dedicated to instrument support?
- Are formal SOPs commonly available?

This snapshot of utility practices and resource commitments is important to understanding drinking water utility attitudes and capabilities with respect to online instruments. A likely use of the survey results is as a rough estimate of the number of utilities that could potentially improve the performance of their online monitoring activities by implementing the best practice SOPs developed in this study. The survey and case study information about instrument application as related to calibration and verification activities offers insight into current industry SOP requirements.

SURVEY METHODS AND RESPONSE

The National Survey of Online Monitoring Practices (appendix A) was mailed in March 1998 to all the utilities represented in the Water:/Stats database. A total of 264 utilities, representing 392 treatment plants, responded to the survey. The overall response rate of the survey was approximately 30%, representing a total of 353 hours dedicated to the preparation of survey responses.

The survey data were entered electronically into a spreadsheet program. The data entry process was monitored for QA and QC to verify the accuracy of data entry.

Database management and analyses were performed in a statistical data analysis program environment. Normalization of survey responses was a critical step in preparing the data for subsequent analyses and included (1) converting text fields to numeric fields for class identifications; (2) converting units of measure to common units (e.g., convert from milligrams per liter to parts per million); and (3) generating appropriate categories of instrument types and location to generalize aspects of the data analysis, as shown in Tables 2.1 and 2.2, respectively.

ONLINE INSTRUMENT AVAILABILITY AND USE

Overall, oxidant analyzers are the highest represented group of instruments among drinking water utilities, where 76% of the respondents indicated that such instruments were used at their plants (Table 2.3). Of those plants having an online oxidant instrument, 91% reported that

Table 2.1
Instrument classes, types, and number of instruments

Analyzer classification (N*)	Analyzer type	No. of instruments (% of class)	
Oxidant analyzers (254)	Chlorine	478	(91)
	Ozone	21	(4)
	Oxidation–reduction potential	10	(2)
	Ammonia	7	(1)
	Chlorine dioxide	4	(<1)
	Potassium permanganate	4	(<1)
Particle analyzers (975)	Turbidity	749	(77)
	Particle count	160	(16)
	Streaming current detector (SCD)	63	(6)
	Surface scatter	3	(<1)
Corrosion control (336)	pH	329	(98)
	Phosphate	4	(1)
	Alkalinity	3	(<1)
Organics control (5)	Gas chromatography	2	(40)
	Total organic carbon (TOC)	2	(40)
	Ultraviolet	1	(20)
General water quality (33)	Conductivity	23	(70)
	Dissolved oxygen	5	(15)
	Dew point	3	(9)
	Color	2	(6)
No instruments		62	

*N = number/count.

they used chlorine residual analyzers (Table 2.1). Oxidant analyzers are closely followed by particle analyzers, reportedly present in 65% of the responding plants. Turbidity online monitors comprised 77% of the particle analyzers reported. Corrosion control instruments are the third most prevalent instrument type (35% of responding plants, with 98% reporting the use of pH analyzers). Next are instruments for monitoring general water quality (16% of responding plants, with 70% reporting the use of conductivity analyzers), followed by instruments for organics control (1% reporting the use of online total organic carbon [TOC] analyzers). Although only five organics control instruments were reported, the authors feel that it is important to maintain this category as part of the evaluation because regulations focus specifically on organics control

Table 2.2
Instrument locations

Location	Example survey responses
Raw	Raw, plant influent, river water, radial-arm collector well
Pretreatment	Presedimentation, clarified, postsedimentation, rapid mix, top of filter, coagulated water, contact basin
Filtered	Filtered, combined filter, granular activated carbon effluent, sand filter effluent
Finished	Tap water, to distribution, treated water, clearwell, combined plant
Other	Carbon dioxide feed, chemical storage site, fluoride, pump effluent

Table 2.3
Number and percent of responding water utilities and treatment plants with types of online instruments installed

	Utilities (N* = 264)		Treatment plants (N = 392)	
Instrument type	No. with instrument type†	% with instrument type†	No. with instrument type†	% with instrument type†
Oxidant analyzer	215	81	296	76
Particle analyzer	203	77	254	65
Corrosion control	128	48	139	35
General water quality	54	20	63	16
Organics control	5	2	4	1
No instruments	41	16	43	11

NOTE: Numbers do not add to 100% because they are a percentage based on a class of instrument.
*N = number/count.
†Most utilities and plants have more than one instrument type installed.

measures. Maintaining this class, then, may be valuable for assessing trends in online instrument use in future projects that deal with online instrumentation. Of the responding plants, 11% indicated that they used no online analytical instruments.

Instrument application information indicates that performance trending is the foremost use of online instruments, accounting for 38% of the overall application category (Table 2.4). Manual chemical feed rate control (23%), compliance reporting (16%), and automated chemical feed control applications (10%) are also significant application types. Plants reported that instruments are used for "other instrument checks" only about 6% of the time. Also, about 6% of the plants reported that instruments are used for applications not specified in the survey.

Table 2.4
Overall indication of online instrument applications by the drinking water industry

Instrument application	Plant responses No.	Plant responses %
Performance trending	1,421	38
Manual chemical feed control	865	23
Compliance reporting	609	16
Automated chemical feed control	359	10
Other instrument check	221	6
Other applications not specified in the survey	235	6
Total	3,710	99*

*The "missing" 1% is a function of rounding.

Table 2.5
Number of plants indicating the application of online monitoring data for classes of instruments

Instrument type	Performance trending	Manual chemical control	Compliance reporting	Automated chemical feed control	Other instrument check	Other applications
Particle analyzer	695	353	239	102	97	71
Oxidant analyzer	367	299	220	114	65	57
General water quality	311	179	126	126	52	40
Corrosion control	233	149	85	80	36	20
Organics control	63	42	40	25	13	12
None	10	6	65	63	65	7

An assessment of application and type was performed to evaluate the relationships between types of online instruments (Table 2.5). This analysis revealed that particle analyzers and oxidant analyzers—used for performance trending, compliance reporting, and manual chemical feed rate control—dominate analytical online instrument use. Further, these instruments are predominantly located at the filtered effluent and finished water locations (Table 2.6) with the finished water location being the most significant. It is clear that turbidimeters at the filtered water location and chlorine residual analyzers at the finished water location are the most prevalent combinations. Note that these data (Tables 2.5 and 2.6) were compiled by assigning a flag to application, type, and location combinations whenever one was recognized. This means that

Table 2.6
Number of plants indicating the use of online monitoring instruments at key locations in the plant

Location	Particle analyzer	Oxidant analyzer	Corrosion control	General water quality
Raw	315	57	106	34
Pretreated	274	169	165	29
Filtered	613	121	49	20
Finished	343	600	228	65
Other	153	178	113	52

instruments could be counted twice if the survey indicated that, for example, a turbidimeter was used for performance trending and compliance reporting.

CALIBRATION AND VERIFICATION ACTIVITIES

An observable difference between calibration and verification information is that verification events generally occurred more frequently than calibration events (Table 2.7). This is to be expected because verification (i.e., checking an online instrument measurement against a bench-top instrument measurement) is typically used as a "quick check" tool between calibration events. The calibration and verification frequency information provided by survey respondents included many records indicating that either calibration or verification was performed. However, no frequency was provided; these are listed in Table 2.7 as "NA." As a possible explanation, many plants may perform these types of instrument checks but not as a regularly scheduled event.

A substantial number of plants did not indicate a frequency of performing verification checks for their online instruments. For those that do practice verification checks, the typical frequencies are weekly to monthly for most online instrument types. Verification checks are performed more frequently than calibration events (Table 2.7), which typically take place on a monthly to quarterly basis for all instrument types. However, a significant portion of the respondents indicated that their online instruments were calibrated only semiannually or annually. Table 2.8 shows the primary calibration and verification frequencies for each instrument type. Refer to Table 2.1 for a description of the instruments included in each instrument class.

Table 2.9 displays the results of evaluating the predominance of different calibration methods by instrument type. Two-point calibration is the primary calibration method for particle analyzers (turbidimeters), corrosion control, and general water quality. Single-point calibration is the most common method for oxidant analyzers.

STANDARD OPERATING PROCEDURES

SOPs were defined in the survey as written documentation that furnished guidance for instrument operation, calibration, and maintenance. The purpose of including a survey of SOP information was to gain some insight about formal SOP use patterns to try to understand if plant operators consider SOPs important for analytical online instrument operation. Interestingly, this

Table 2.7
Calibration and verification frequency indicated by responding plants

Instrument type	Frequency	No. of plants with calibration frequency	No. of plants with verification frequency
Particle analyzer	Daily	18	88
	Weekly	65	246
	Monthly	201	246
	Quarterly	310	1
	Semiannually	111	0
	Annually	124	0
	NA*	139	394
Oxidant analyzer	Daily	41	75
	Weekly	35	178
	Monthly	137	123
	Quarterly	124	0
	Semiannually	44	0
	Annually	43	0
	NA	103	151
Corrosion control	Daily	4	41
	Weekly	28	99
	Monthly	143	83
	Quarterly	83	0
	Semiannually	23	0
	Annually	23	0
	NA	32	113
General water quality	Daily	1	6
	Weekly	5	23
	Monthly	24	20
	Quarterly	18	0
	Semiannually	5	0
	Annually	33	0
	NA	23	60
Organics control	Daily	1	1
	Weekly	0	0
	Monthly	2	0
	Quarterly	1	0
	Semiannually	0	0
	Annually	0	0
	NA	1	4

*NA = No frequency provided.

Table 2.8
Most frequently reported frequencies (mode) for calibration and verification activities by survey respondents

Instrument type	Calibration frequency	Verification frequency
Particle analyzer	Quarterly	Weekly
Oxidant analyzer	Monthly	Weekly
Corrosion control	Monthly	Weekly
General water quality	Annually	Weekly
Organics control	NA*	NA

*NA = No frequency provided.

Table 2.9
Calibration methods used for each instrument type

Instrument type	Calibration method	No. of responses
Particle analyzer	1-point	425
	2-point	750
	Multiple	9
	Other	244
	No method specified	290
Oxidant analyzer	1-point	401
	2-point	359
	Multiple	21
	Other	175
	No method specified	172
Corrosion control	1-point	52
	2-point	422
	Multiple	15
	Other	62
	No method specified	52
General water quality	1-point	49
	2-point	90
	Multiple	11
	Other	21
	No method specified	42

Table 2.10
Plants with and without SOPs by instrument type

Instrument type	Total No.	% of plants with SOPs	% of plants without SOPs
Particle analyzer	960	91	9
Corrosion control	334	93	7
Oxidant analyzer	526	93	7
General water quality	108	79	21
Organics control	5	—	—

survey indicates that most plants have formal SOPs in place for online instruments, suggesting that they must be considered integral to instrument use (Table 2.10). More than 90% of the responding plants have SOPs for particle analyzers, corrosion control, and oxidant analyzer instruments.

CONCLUSIONS

The majority of utilities responding to the national survey indicated that they used some form of online monitoring instrument (84%). The two major types of online monitoring instruments used by utilities were turbidimeters and chlorine analyzers; however, many utilities clearly use a range of online instruments for routine water quality monitoring. Further, online instruments within all classes were found to be located throughout plants, with predominance determined by the likely uses of the data. For example, the majority of online turbidimeters were found to be located at filter effluents and the majority of chlorine residual analyzers were found to be located at the finished water.

The application of online monitoring data showed for all instrument types that the prevalence of uses by utilities is ranked as follows:

> Performance trending > manual chemical feed controls > compliance reporting > automated chemical feed controls > other instrument checks > other applications

With these uses of online monitoring data, the majority of utilities indicated that instrument calibration was performed at least quarterly. Field verification was performed at least monthly (and weekly in many cases). Clearly, as the motivation for using online monitoring data for such applications as compliance reporting and automated control of plant operations strengthens, utilities may increase the frequency of calibration and verification checks to accommodate the higher data quality objectives associated with those applications.

CHAPTER 3
CASE STUDY RESULTS

The performance of online instruments at six case study utilities (designated as A–F) was evaluated in this study with a goal of gaining field-based insights into operational issues, applications of measurement data, and utility support requirements for these instruments. A range of instrument types were included among the case study utilities, although common to at least half of the participating utilities were the use of online turbidimeters, chlorine analyzers, and pH meters. The online instruments also represented a variety of manufacturers, and this study did not distinguish online instruments on the basis of either model or manufacturer type. Instead, the objective of the case studies was to encompass the array of instruments that utilities are currently using and to evaluate these instruments in the field. Newer online instruments are also included, but the comparative evaluation of these instruments among case study utilities was not feasible because of the low numbers of such applications.

ONLINE MONITORING APPLICATIONS AT CASE STUDY UTILITIES

For each case study utility, a site visit was conducted to obtain descriptive information about the utility's online instrument use, current practices with respect to online instrument support and maintenance, and SOPs. Appendix B of this report contains a detailed summary description of each of the case study utilities.

The case study utilities included surface water systems that used various technologies and relied on various source water conditions (Table 3.1). The case studies encompassed a range of water quality conditions, and the treatment challenges of these facilities were significant in terms of the online monitoring program support currently provided at the utilities (Table 3.2). For example, at surface water plants, online turbidimeters are used on both the individual filter and combined filter effluents for each case study that operated a filter plant. Two of the case studies used online turbidimeters for the filter backwash streams; two others used online turbidimeters at intermediate treatment steps preceding the filters.

In general, the case study utilities performed online monitoring to obtain trending information for manual operation of the water treatment plant. However, several more specialized applications were also available through the case study utilities:

- Compliance reporting for such requirements as the monthly turbidity and chlorine residual measurements set forth in the Surface Water Treatment Rule
- Automatic control of backwash operations
- Automatic control of chemical feed applications (such as acid addition during coagulation, coagulant dose, carbon dioxide, sodium hydroxide, ozone, phosphate inhibitors, and fluoride)
- Early warning signals for source water operations (blending, intake shutdown, reservoir turn-over, high oxidant demand)
- Chemical feed storage volumes
- Research applications for future treatment applications and automated control to optimize treatment performance objectives

Table 3.1
Characteristics of case study utilities

			Average (range) source water quality				
Case study utility*	Treatment plant type	Source water type	Turbidity (ntu)	Alkalinity (mg/L)	pH (SU†)	Temperature (°C)	TOC (mg/L)
A	Conventional	Reservoir	4.8 (0.3–24)	97 (70–120)	7.7 (7.3–8.2)	16 (10–22)	4.4 (3.6–5.0)
	Direct filtration	Reservoir	0.9 (0.3–1.9)	15 (10–21)	7.3 (6.4–8.2)	15 (10–19)	1.1 (0.9–1.5)
	Direct filtration	Reservoir	5.8 (0.8–42)	77 (59–110)	7.5 (7.1–7.9)	15 (10–21)	3.8 (3.3–4.7)
B	Direct filtration	River with seasonal groundwater	3.4 (0.8–8.5)	82 (68–118)	7.9 (7.6–8.3)	12 (4–20)	3.0 (2.2–4.0)
C	Lime softening	River with groundwater	2.6 (0.7–5.5)	208 (195–215)	8.1 (7.9–8.2)	16 (7–24)	6.0 (4.0–7.7)
D	Conventional filtration	River	7.7 (2.2–15.5)	73 (57–90)	8.0 (7.6–9.2)	17 (4–28)	2.5 (1.5–3.5)
E	Disinfected, unfiltered	Reservoir and groundwater	1.8 (0.3–6.6)	16 (14–20)	7.2 (7.0–7.4)	11 (7–22)	1.0 (<0.7–1.6)
F	Conventional filtration	Lake	3.3 (0.4–40)	120 (111–140)	8.4 (7.8–8.8)	11 (<1–27)	2.0 (1.8–2.4)

*Case studies are differentiated by the letters A through F.
†SU = standard units.

Table 3.2
Online instrument applications at case study utilities

Online instrument type	Online monitoring location			
	Raw water	Process water	Finished water	Distribution system
Alkalinity	F			
Ammonia			A	
Chlorine		A,B,C,D,F	A,B,C	E,F
Fluoride		F	A	E
Ozone		A,C		
Particle count	B	A,C,D	B	
pH	A,D	A,C,D		E
Phosphate		G		
SCD		D		
Temperature	F			
Turbidity	B,C,E,F	A,B,C,D,F	C,D	
UV_{254}*			C†	

*Ultraviolet at a wavelength of 254 nanometers.
†Pilot-plant application.

The support of online instruments at the case study utilities exhibited many consistencies as well as some interesting differences (Table 3.3). The fact that these utilities are all classified as large systems explains why most of the case study utilities have dedicated instrumentation personnel. Usually this position was either in the operations, laboratory, or water quality departments. It was also common to see different departments being responsible for different aspects. At one utility, a team approach was employed where three individuals supplied different aspects of the support, testing, operation, and calibration and maintenance. Another unique event that quite a few of the case study utilities mentioned is side-by-side testing. This testing entails installation of two or more similar instruments (that measure the same parameter) so that direct comparisons of performance can be conducted. A process engineer typically directed this type of testing.

One of the utilities practices a unique approach to instrument support. Operations addresses general maintenance and quality checks on an as-needed basis. However, instrumentation technicians, who are assigned specific instruments, perform all the detailed troubleshooting and calibration. This particular utility strongly believes that this approach creates a sense of ownership and that, as a result, their instrument support is superior.

Many of the case study utilities have SOPs in place. However, a few had no SOPs for their online instruments at all. Even though SOPs range widely in the information they contain, calibration methods and frequencies were common to all the SOPs reviewed. In some cases the manufacturer's procedures were directly referenced and attached; in other cases, the SOP is

Table 3.3
Summary of case study utility online instruments evaluated by the monitoring program

		Online instrument type						
Case study	Location	Turbidity	Chlorine	Fluoride	pH	Phosphate	Alkalinity	Ozone
A	Raw	×						
	Applied	×						
	Filter effluent	×						
	Treated	×						
B	Raw	×						
	Filter effluent	×						
	Distribution	×	×					
C	Raw	×						
	Recarbonation				×			
	Postozone							×
	Filter effluent	×						
	Clearwell				×			
	Pilot plant							
D	Rapid mix				×			
	Filter influent		×		×			
	Filter effluent	×	×					
E	Raw	×						
	Finished			×				
F	Raw						×	
	Finished					×		

presumed to be a hybrid of the manufacturer's calibration procedure and plant-specific protocols. Personnel at the utilities that had SOPs felt that they were worthwhile and reported having spent considerable time developing them. The few utilities that did not have SOPs also felt that they would be worthwhile, but generally cited time and labor constraints as the limiting factors.

DATA COLLECTION AND ANALYSIS METHODOLOGY

Several of the participating utilities collected paired samples of bench and online measurements for the instruments included in this study as shown in Table 3.4 for the various instruments included in the study. For all case study utilities, this data collection effort was

Table 3.4
Overall performance of turbidimeters in terms of agreement with bench measurements

Attribute	Raw water turbidimeters	Filtered water turbidimeters
Bench turbidity measurements (ntu)		
Minimum	0.06	<0.001
Average ± standard deviation	2.7 ± 2.1	0.080 ± 0.040
Maximum	47.5	4.0
Delta (online – bench) (ntu)		
Average ± standard deviation	–0.37 ± 0.95	–0.011 ± 0.084
Relative delta (delta/bench) (%)		
Average ± standard deviation	–5 ± 54	–4 ± 58

performed for approximately a 1-year period. Additional information was gathered from some case studies based on historical data collection efforts that the utility had performed. These data were integrated into the analysis performed for this study when appropriate.

The data collected were organized into a series of databases that could be easily merged and manipulated for graphical and statistical analyses. Maintenance and calibration events were noted by the case study utilities for the days on which they occurred. These events were converted into a series of "flags" in the database (binary codes of "0" and "1" by event type to signal the timing of such events). The trending of the relative performance between bench and online instruments was evaluated with respect to such calibration and maintenance events. In addition, these events were catalogued, with estimates of utility support requirements generated.

Several terms were developed in the course of the data analysis and reduction. The difference between online and bench measurements (calculated as [online] – [bench]) is termed the "delta," and indicates the degree—or lack thereof—of agreement between bench and online instrument measures. The running average delta is an average of 30 sequential delta estimates, and changes in this term may reflect changes in the degree of agreement between online and bench-instrument measures over time. The "volatility" of online measurements was also assessed, calculated as the running variance (30 sequential measurements used to estimate the variance) in online measurements. This measure reflects the sensitivity of online measurement trends to unusual values, potentially measurements that are aberrant (outliers) or that reflect a short-term deviation in water quality conditions. The measure can also represent the precision of the online instrument measurements when the "true" variance of a water quality constituent has been estimated a priori using reliable measurement techniques. In other words, the running average delta and variance can serve as indicators of the degradation in online instrument measurement performance over time (Figure 3.1).

This study evaluated the performance of online instruments as compared to bench instrument measurements at six case study utilities. The objective of this analysis was to assist utilities in understanding the differences they can expect between online and bench measurements and how those differences can ultimately affect the application of such data.

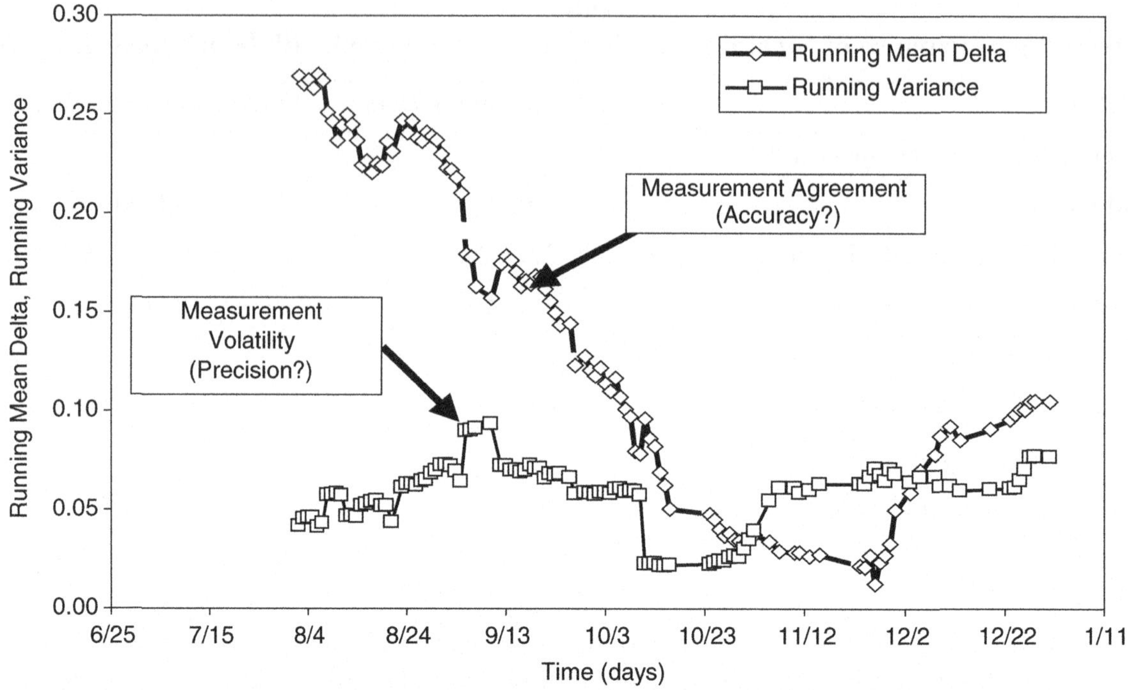

Figure 3.1 Different measures of online measurement accuracy and precision that may be useful for utilities to incorporate in their routine operations

The routine bench and online measurements were compared graphically for all instruments included in the six case studies. The differences between bench and online measurements varied substantially among the case study utilities and among the instrument types. As illustrated in Figure 3.2, some instruments experienced a clear bias in the comparative measures where the online measurements were either consistently higher or lower than the online instrument measurements. In this example, the online chlorine measurements were consistently lower than the bench measurements, with some notable exceptions (Figure 3.3). The occasional but large overestimations of chlorine residual by the online analyzer may indicate an important deviation in typical analyzer performance. Identifying such anomalies in online instrument measurement performance is discussed later in this chapter.

Bench and online measurements should be linearly related because they are measuring the same water quality condition. Of course, differences are expected, but these differences are expected to relate either to (1) a bias in instrument measurements between bench and online results, or (2) a random error in measurement duplication. When a bias in measurements occurs between bench and online instruments, the linear relationship between these results should have a slope significantly different than 1, as was the case in the chlorine analyzer example for case study D (Figure 3.4). However, if the measurement difference represents random error, the linear relationship should have a slope equivalent to 1. In either case, the goodness-of-fit for the linear relationship is an indicator of the variability between bench and online measurements. In this case, the authors have used the correlation coefficient, r^2, to indicate the goodness-of-fit. For the online and bench chlorine measurements collected by case study D, the r^2 is fairly poor given that these two instruments were basically measuring split samples. The variability exhibited by these

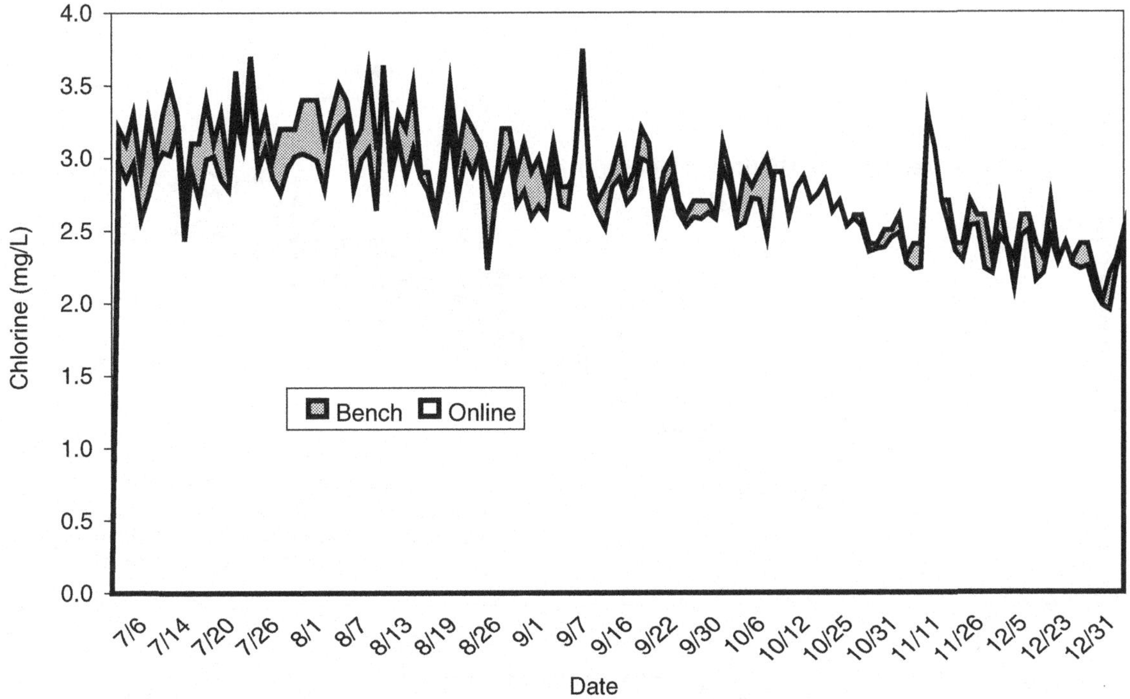

Figure 3.2 Example comparison of bench and online chlorine measurements (case study D). Systematic bias in measurements is observed in this case where the bench measurements are predominantly higher than the online measurements.

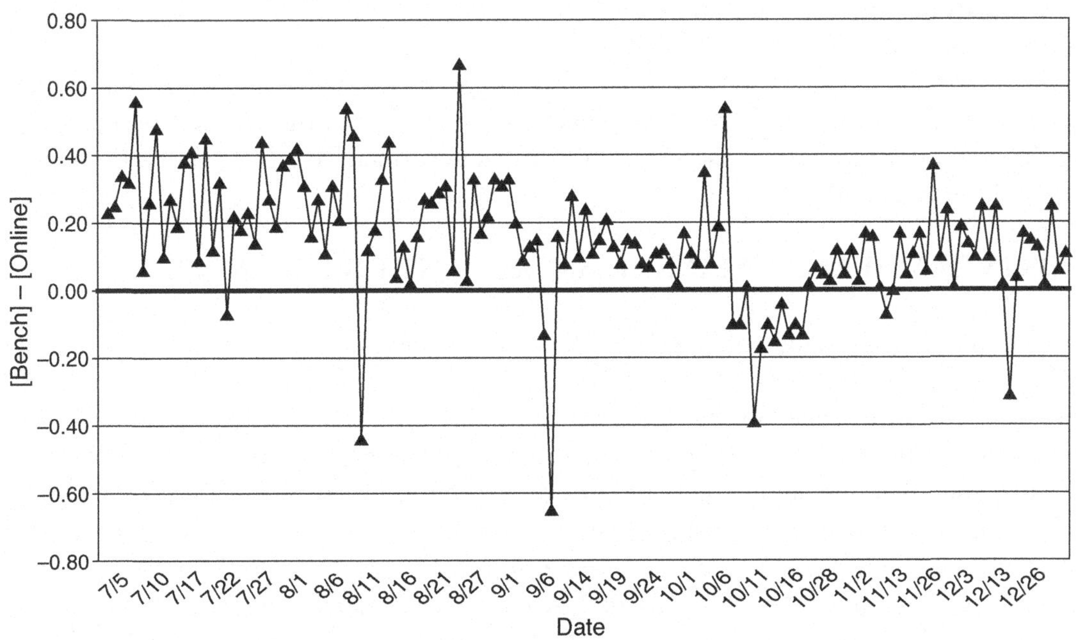

Figure 3.3 Difference in bench and online chlorine measurements (delta) for case study D

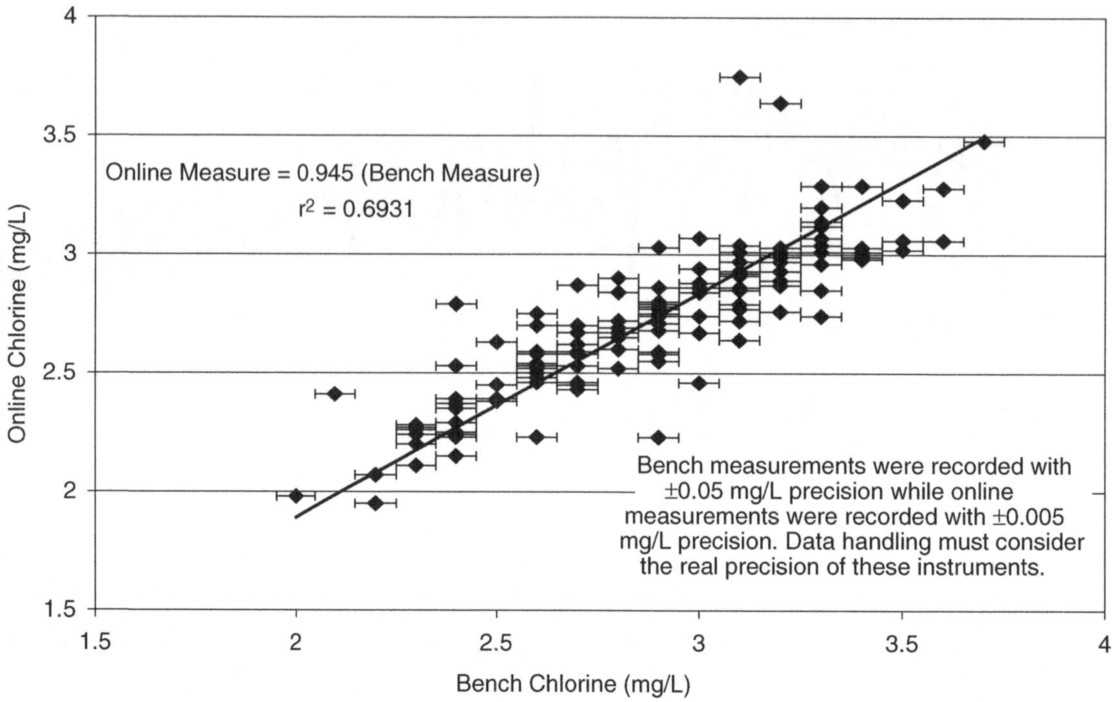

Figure 3.4 Linear relationship between bench and online chlorine measurements with precision estimates shown

instruments is beyond either the accuracy or precision issues and represents true deviations (or errors) in measurement. Determining which instrument has a measurement error issue would be an important followup task for this utility.

In the next chapter, the issues related to online data quality are addressed directly. The results presented in this chapter will relate to the observation of the online and bench measurements collected by the case study utilities as described above for the various instrument types included in the study.

TRENDS IN BENCH AND ONLINE MEASUREMENT DIFFERENCES

The monitoring data collected by the various case study utilities were evaluated to determine trends in differences between bench and online measurements. The term delta is used to represent this difference and is calculated as the online measurement minus the bench measurement. Delta indicates the bias in measurement differences between online and bench instruments, such that an average delta of zero shows that no bias exists; a nonzero average delta indicates a directional difference in bench and online measures. The term "relative delta" is used to mean the relative proportion of the bench measurement represented by the delta (expressed in percent) and is calculated as the delta divided by the actual bench measurement.

Several of the case study utilities evaluated online turbidimeters and chlorine analyzers, and, in some cases, multiple instruments were included within the same plant. Therefore, a larger population of data was available for these instruments, meriting the more in-depth analysis and discussion that follows. For the remaining instruments evaluated in the course of the study, little

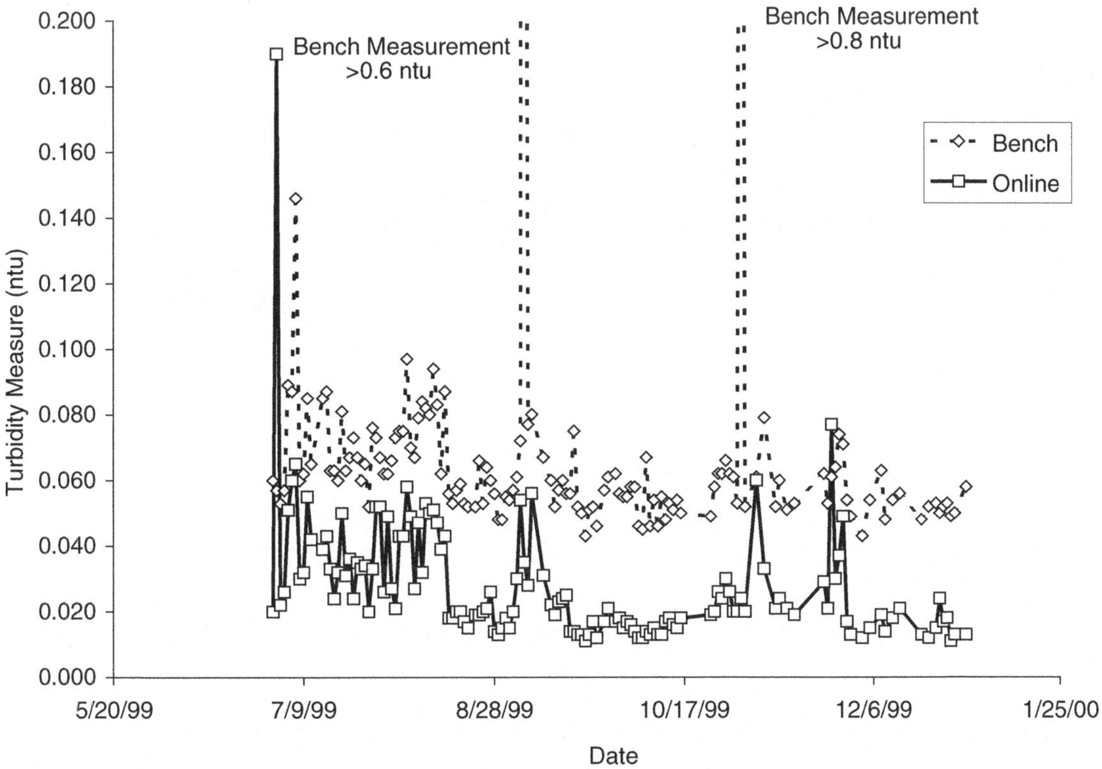

Figure 3.5 Time series of bench and online finished water turbidity measurements for case study E

replication was available, and so the results are presented for each instrument type evaluated in a more abbreviated manner than that possible for turbidimeters.

Online Versus Bench Measurements for Turbidimeters

Among the case study utilities, only turbidimeters were included in the comparison of bench and online measures. The difficulties associated with comparability, standardization, and calibration of particle count instruments have been studied elsewhere (VanGelder, Chowdhury, and Lawler 1999).Therefore, this study focused on the use of online turbidimeters, because these instruments are critical to surface water plant operations. Four case study utilities evaluated turbidimeters located at the raw water supply and for filtered water (either individual filters or combined filter effluents). For case study A, 10 online turbidimeters measuring individual filter effluents were evaluated separately for their interinstrument performance. This analysis is discussed separately later in this chapter. The bench and online turbidity samples were collected as paired samples, intended to represent the same water quality condition, over a period of 1 year (Figure 3.5). To minimize the impact on the participating utilities, the standard practices with respect to the frequency of bench measurements was maintained during the course of the study. Therefore, the number of observations per case study utility and per instrument location varied. In no case, however, were less-than-weekly samples collected and evaluated.

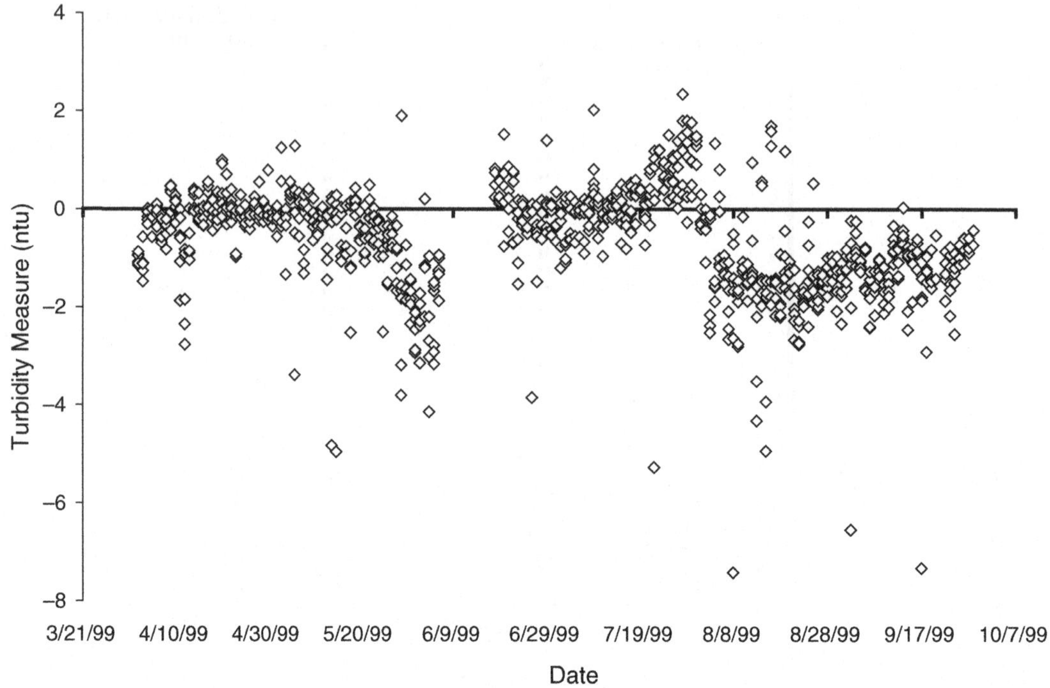

Figure 3.6 Time series of delta values (online – bench) raw water turbidity measures for case study B

The online turbidimeters demonstrated an overall bias in their measurements as compared to bench measurements used at the case study utilities. This is clearly illustrated for case study E where the online measurements are consistently and substantially lower than those found through bench measurement (Figure 3.5). Case study B collected a substantial number of comparative turbidity data, with the utility reporting its monitoring results every 4 hours for both raw and filtered turbidity measurements. As a result, this case study provides an in-depth data set of observations from which the correspondence of bench and online measures can be derived. As shown in Figures 3.6 and 3.7, differences between bench and online turbidity measurements occurred over time, and the online turbidimeters were generally found to have lower measurements than the bench instruments. By regressing the online measurements as a function of bench measurements, the slope of the best-fit line was 0.80 for raw water measurements and 0.93 for filtered water measurements, demonstrating the underestimation of turbidity by online instruments as compared to bench instruments (Figures 3.8 and 3.9). However, this does not imply that the online measures are more or less correct than those found using the bench instruments—it simply means that different results will be obtained depending on whether an online or bench instrument is used. Also, in general, the online instruments are likely to find lower turbidity levels than bench instruments.

For each of the online turbidimeters evaluated, the overall mean and standard deviation of delta were found (Figure 3.10). Raw water instruments clearly showed the greatest magnitude of delta and its variability. However, when the delta values were normalized by the actual measurement (bench) level, the relative mean delta and the standard deviation of relative delta

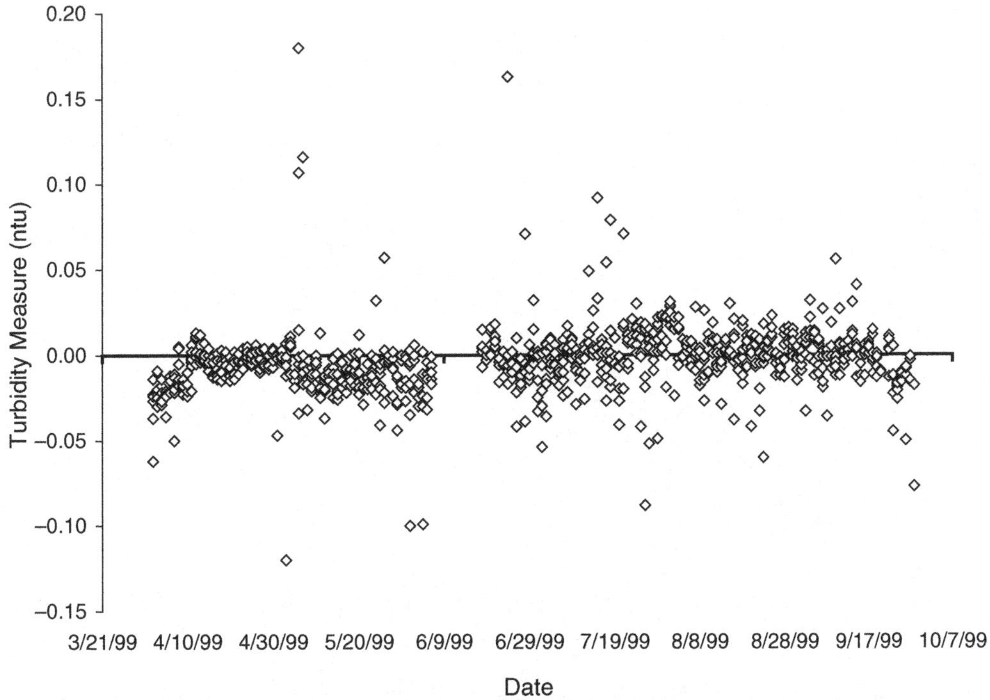

Figure 3.7 Time series of delta values (online – bench) for finished water turbidity measurements for case study B

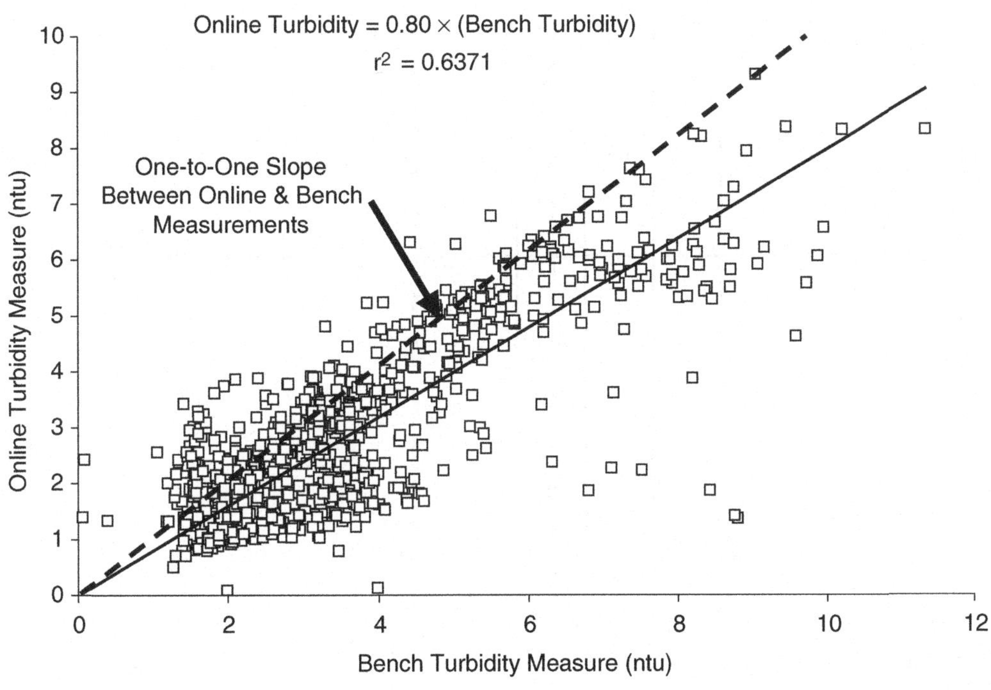

Figure 3.8 Correspondence between bench and online raw water turbidity measures for case study B

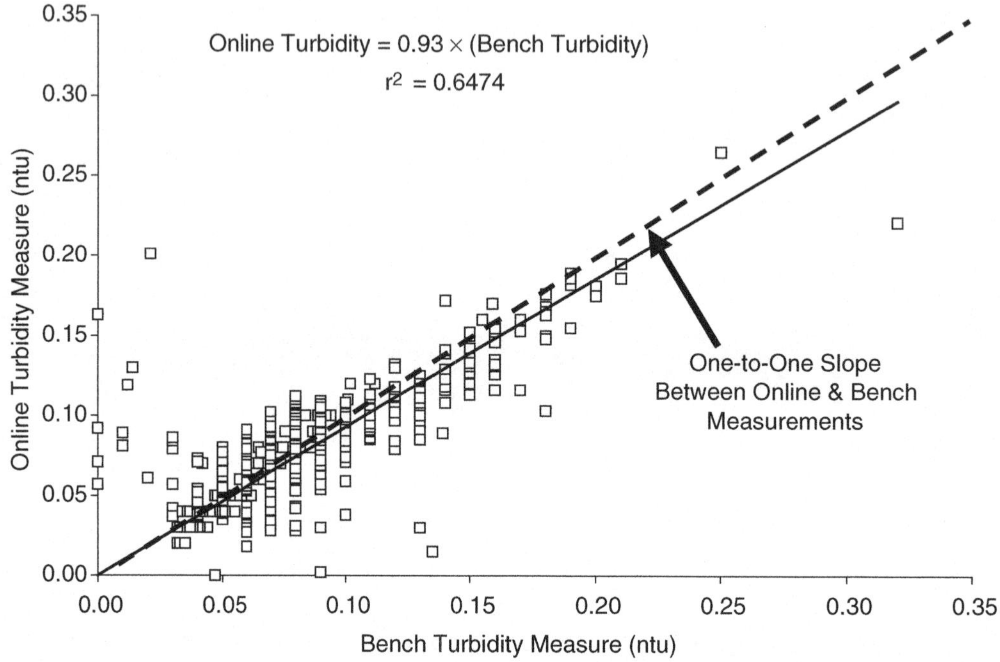

Figure 3.9 Correspondence of bench and online filtered water turbidity measures for case study B

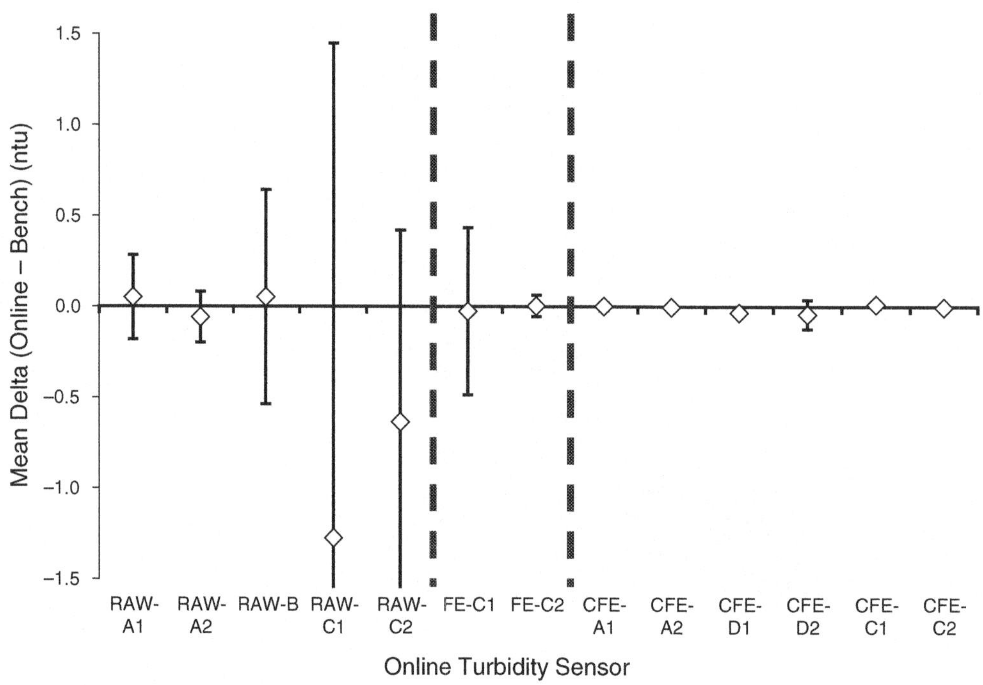

NOTE: FE = filter effluent; CFE = combined filter effluent.

Figure 3.10 Overall mean delta for turbidity measurements for all case study instruments evaluated based on monitoring location

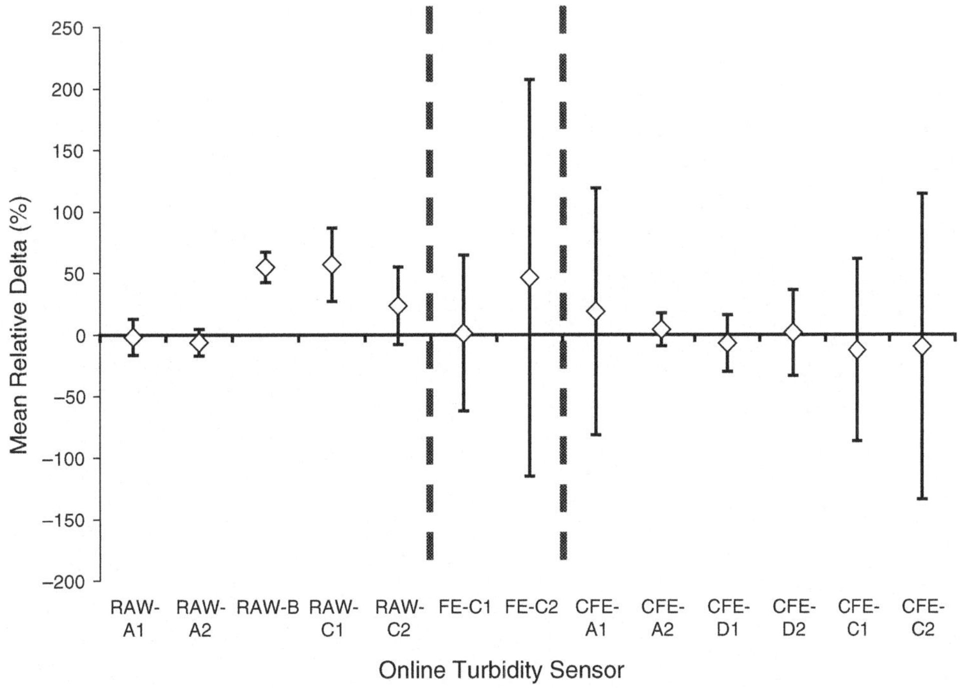

Figure 3.11 Overall mean relative delta for turbidity measurements for all case study instruments evaluated based on monitoring location

values were found to be greater for filtered water online measures than for raw water measures (Figure 3.11). The very low turbidity levels found in filtered water contribute to large relative disagreements when small delta values are seen. Inversely, the larger delta values for raw water measurements represent only a small proportion of the actual measurement because of the higher values for raw water turbidity.

Clearly, the location of the online monitoring application affects the apparent performance of such instruments as compared to bench measurements. When aggregated for all instruments included in the study, the average delta value for raw water measures was –0.37 ntu (±0.58 ntu standard deviation), and the average delta value for filtered water measures was –0.011 ntu (±0.020 ntu standard deviation; Table 3.4). In terms of relative delta values, the average relative difference between online and bench measurements was comparable regardless of location (–5% for raw water and –4% for filtered water). The variability in relative delta measures for filtered water applications was also comparable to that found for raw water measures (standard deviation of relative delta for filtered water was 58% versus 54% for raw water measures). These findings suggest, then, that (1) online turbidity measurements may be biased low as compared to bench measurements for turbidity; and (2) the variability in online and bench comparative measurements is very high, suggesting an instability in measurement performance among bench and online instruments.

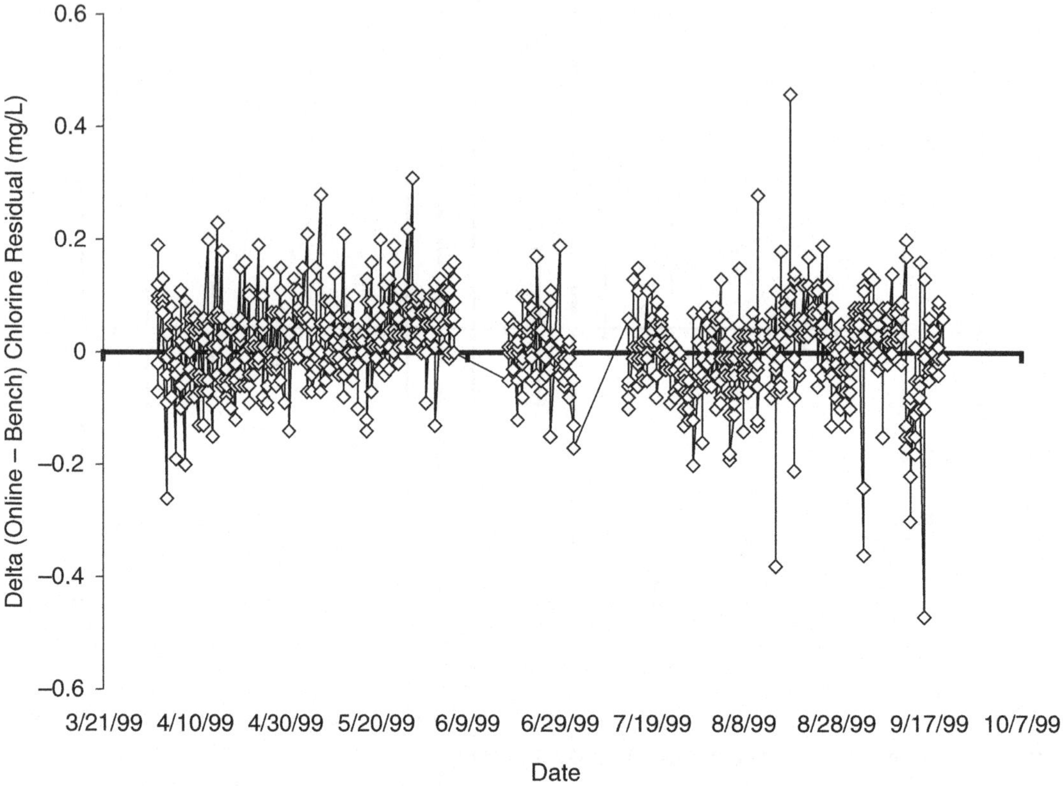

Figure 3.12 Time series of delta chlorine residual measurements (online – bench) for case study B finished water chlorine measurement

Online Versus Bench Measurements for Chlorine Analyzers

Two of the case study utilities evaluated several chlorine analyzers. In general, the chlorine analyzers experienced a narrow range of measurement values (within ±1 mg/L of chlorine from minimum to maximum measurements) and the levels were consistently within the detection range of both the bench and online instruments. In some cases, however, the variability between bench and online measurements over time was considerable. The delta values ranged from nearly 0 to ±0.4 mg/L in the example shown in Figure 3.12. Although little to no bias was found in some cases (slope of 1.0), the degree of variability between online and bench measurements led to poor goodness-of-fit when linear relationships were evaluated for these measurements ($r^2 = 0.256$; Figure 3.13).

In general, chlorine analyzer performance was less sensitive to location than turbidimeters, although overall differences were observed when comparing process water measurements of chlorine residual to finished water measurements (Figure 3.14). The mean delta value for the various instruments located at process water locations appeared to have a distinguishable bias when compared to the paired bench samples. In three of the four analyzers, the online chlorine instrument produced lower measurements of residual than the bench instrument. For the analyzers measuring finished water chlorine, the mean delta values were nearly zero in all three cases. The magnitude of the variability in delta values—standard deviation about the mean delta value—was comparable between process and finished water locations for

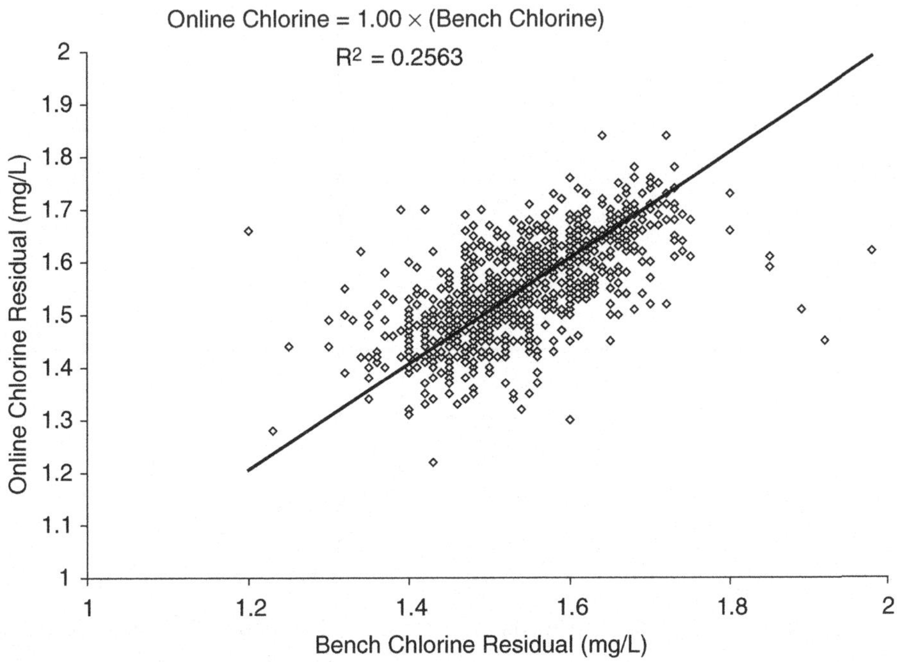

Figure 3.13 Correspondence of bench and online chlorine residual measurements for finished water at case study B

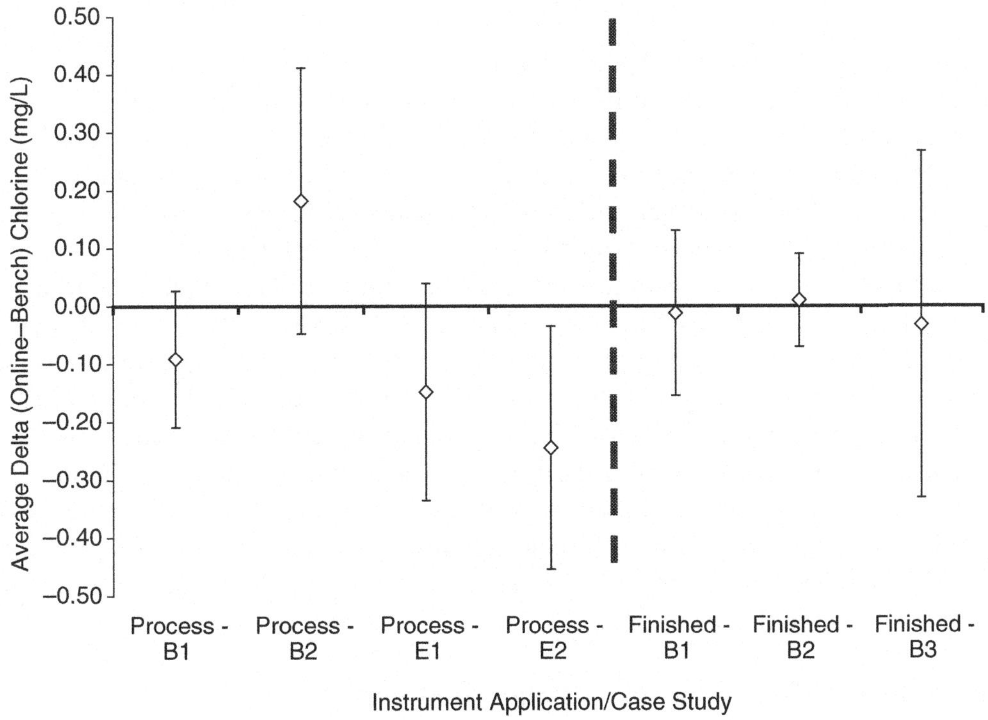

Figure 3.14 Overall mean delta for chlorine residual measurements for all case study instruments evaluated based on monitoring location

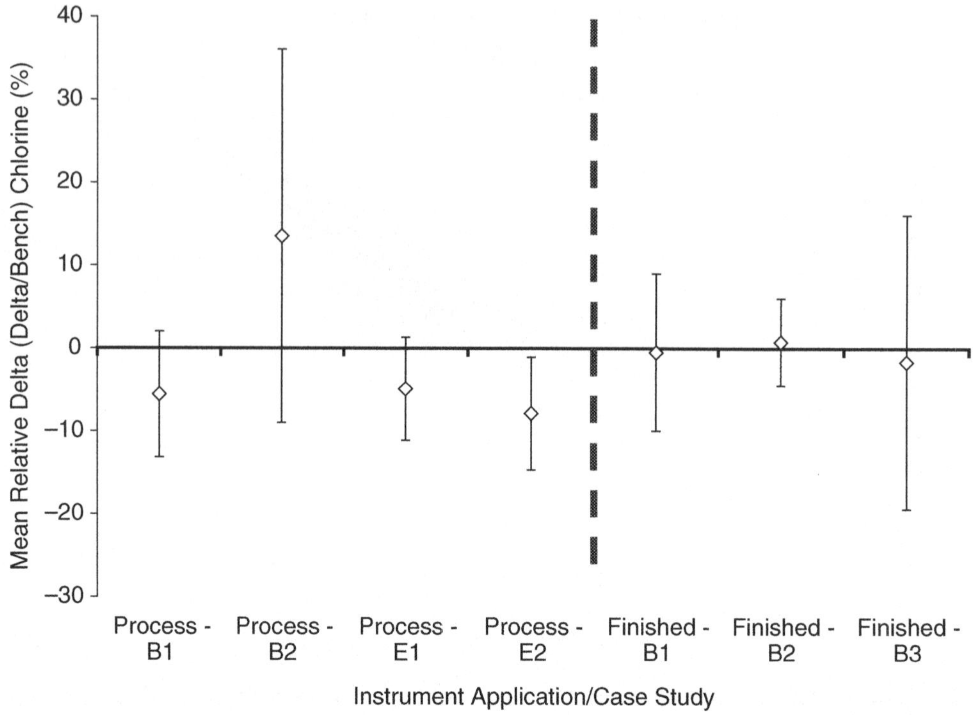

Figure 3.15 Overall mean relative delta for chlorine residual measurements for all case study instruments evaluated based on monitoring location

chlorine residual measurements. As shown in Figure 3.15, the mean relative delta was approximately –5% ± 7% for three of the four online chlorine analyzers used to measure process water chlorine levels. The remaining analyzer had a mean relative delta of +10% ± 22%, indicating significantly different performance than the other three analyzers. Indeed, this analyzer might be characterized as an outlier, and its performance difference from the others might be considered as a trigger for further investigation of instrument performance.

The finished water measurements showed similar performance among the three instruments in terms of the mean delta and mean relative delta levels. However, the variability about the mean difference between online and bench measurements was quite different for the three instruments—approximately 5%, 10%, and 20%, respectively, for the various instruments in terms of the relative delta levels. Table 3.5 summarizes the overall performance among the online chlorine analyzers. Measurements from the process water chlorine analyzers had a clear bias as compared to bench measurements, showing lower levels of residuals than the bench instruments. Certainly, process water applications of chlorine represent a more unstable environment for chlorine residual, and the measurement differences may result from the water quality matrix conditions encountered in this location.

Online Versus Bench Measurements for Other Analyzers

The other online analyzers evaluated in the course of this study included instruments for measuring phosphate, alkalinity, fluoride, and pH. Case study F uses the online measure of phosphate to control corrosion inhibitor dosages. A secondary benefit was observed when the utility

Table 3.5
Overall performance of chlorine residual analyzers in terms of agreement with bench measurements

Attribute	Process water	Finished water
Bench chlorine residual measures (mg/L)		
Minimum	0.040	0.70
Average ± standard deviation	2.2 ± 0.84	1.6 ± 0.088
Maximum	3.8	2.7
Delta (online – bench) (mg/L)		
Average ± standard deviation	–0.16 ± 0.17	–0.011 ± 0.17
Relative delta (delta/bench) (%)		
Average ± standard deviation	–6.1 ± 7	–0.4 ± 11

experienced a break in its chemical feed line detected as reduced phosphate levels. The phosphate analyzer is located in the finished water. Alkalinity is measured online to determine coagulation requirements for case study F because changes in alkalinity are linked to changes in coagulation conditions for the raw source water. This analyzer is therefore located where it can measure raw water alkalinity levels. Case study E, an unfiltered surface water supply, studied the online measurement of fluoride for chemical feed control and pH for corrosion control requirements.

Table 3.6 summarizes the results for the various online analyzers. In all cases, the mean delta between online and bench measurements was negative, indicating that the online measurements were lower than those given by bench instruments. Fluoride and pH analyzers had the greatest relative variation in differences between online and bench measurements. In contrast, phosphate and alkalinity online instruments had good agreement and relatively low variability in differences between online and bench measurements. This should be qualified by the limited evaluation of analyzers. In addition—at least for alkalinity—the study period did not include the seasonal effect of dramatic changes in alkalinity experienced during typical years.

ONLINE MEASURE VOLATILITY

The variability in online instrument measurements over time is important to the application of these data in making operational decisions as well as improving a utility's understanding of its water quality conditions. In some cases, the variability represented by online measurements represents true changes in water quality. However, a portion of the variability may be explained by instrument measurement error. Distinguishing "true" water quality variance from that introduced by measurement error is critical to the application of online measurement data. For example, plants that control chemical feed rates on the basis of an online instrument measurement could see inappropriate chemical dosing conditions when measurement errors occur. "Dampening" the online data stream is often practiced in these cases so that (1) no single instrument signal could result in a major treatment upset event, or (2) the chemical feed pacing

Table 3.6
Overall performance of various online analyzers in terms of agreement with bench measurements

Attribute	Phosphate ($mg/L\ PO_4$)	Alkalinity ($mg/L\ CaCO_3$)	Fluoride (mg/L)	pH (SU)
Bench measures (mg/L)				
Minimum	0.67	110	0.49	6.22
Average ± standard deviation	1.02 ± 0.10	116 ± 3	0.75 ± 0.13	7.12 ± 0.45
Maximum	1.05	128	0.94	7.96
Delta (online – bench) (ntu)				
Average ± standard deviation	–0.050 ± 0.065	–0.3 ± 2.9	–0.04 ± 0.12	–0.22 ± 1.2
Relative delta (delta/bench) (%)				
Average ± standard deviation	–5 ± 6	–0.3 ± 3	–5 ± 15	–3 ± 17

would not continuously search for the correct setting, resulting in highly variable chemical feed rates and prematurely damaged feed pump motors.

Several different methods of dampening online measurements are available and utilities have used these methods to support various applications of online measurements:

- Measurement bandwidths or ranges. This technique sets a series of measurement bands within which the operational system responds within a defined manner (e.g., the dosage levels are related to the measurement band and operator alarms are tripped when outside a certain measurement band).
- Running or moving averages. This method uses a running (or moving) average measurement result instead of the instantaneous measurement. In this case, the measured result collected at each time step is averaged with a defined number of preceding measurements. For example, if data are collected every 5 minutes, the 29 prior measurements and the newest measurement are averaged to generate a running average for that latest 5-minute interval. In this way, the data representing a 2.5-hour window of time are used as the basis for setting operational responses. The time interval represented in the running average measurement may be linked to the expected lag time between a measurement response and the operational action being signaled. For example, if chlorine dose is controlled by chlorine residual measurements and there is a 1-hour lag time from the point of chlorine application and residual measurement, the running average condition used to signal a change in dose should be set to at least a 1-hour interval. The online sensor cannot detect the change in the chlorine dose before this interval.
- Integrated measurements. With this technique, an integrated result is sent within a defined time step. For example, an instrument can send a signal every 30 seconds, but the SCADA system is set to respond to the average of those 30-second signals over a 15-minute data collection period (e.g., 30 measurements are averaged to produce a single measurement every 15 minutes).

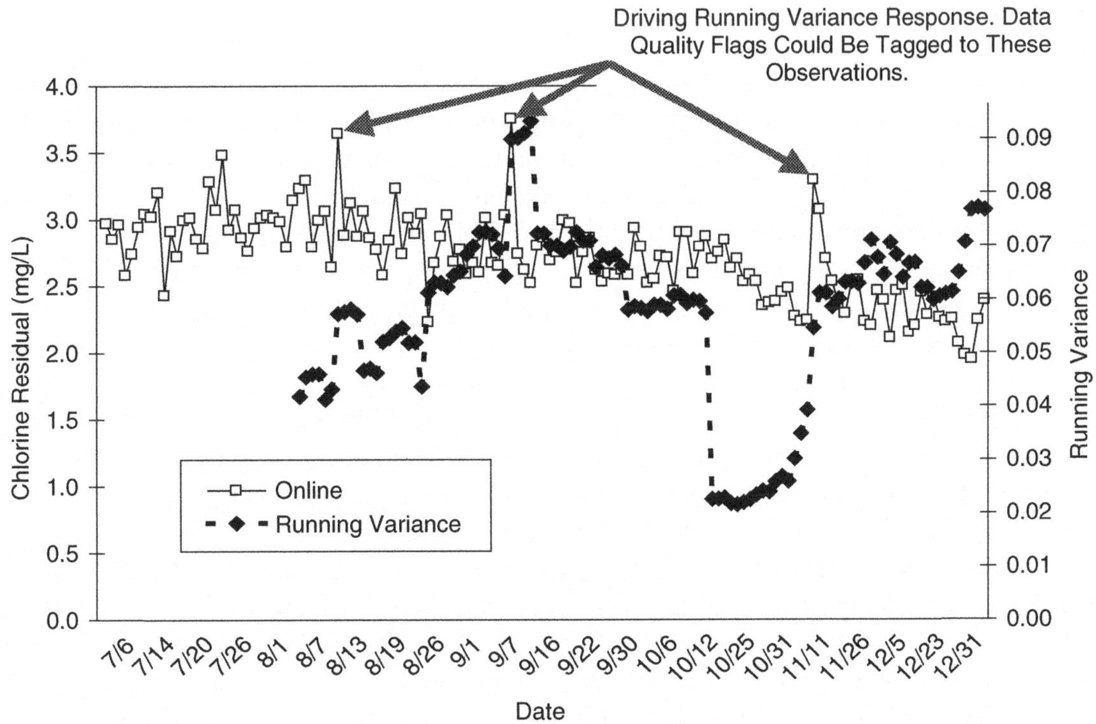

Figure 3.16 Example of online instrument measurement volatility and how it is affected by individual measurement results. These results can be flagged as questionable observations, and the frequency of such flags could be used to signal instrument maintenance or calibration requirements.

Any combination of the above dampening methods (or others) can be used together or individually to address problems associated with automating operational decision making based on online instrument measurements.

However, these methods do not directly address the issue of measurement error—instead, they mitigate the impact of instrument measurement error only. To support online instruments and optimize the range of data use, determining the "quality" of the data produced by the online instruments is important. The measure of volatility created in this study may help utilities to identify aberrant measurements and flag these measurements as either (1) questionable data for inclusion in water quality evaluations or (2) possible indicators of instrument support being necessary (Figure 3.16). The richness of online data sets is a valuable resource that allows utilities to better understand plant operations and water quality conditions. Trending such data over time or performing comparative assessments of plant performance from year to year (or month to month, etc.) can yield important insights that can be used to improve both operations and water quality. By flagging aberrant measurements for possible exclusion in such analyses, more appropriate measures of both plant performance and water quality conditions can be produced. In addition, these flags could be related to instrument support requirements, and, therefore, be used to signal the need for special maintenance or calibration requirements.

CONCLUSIONS

The use of online analyzers at the case study utilities resulted in a variety of uses and applications for supporting treatment operations, operational decisions, trending, and water quality and treatment performance characterization. Each of the utilities used some type of online turbidity analyzer, and nearly all used online chlorine residual analyzers within their systems. Several case studies applied online instruments within the distribution system as well as to measure water quality conditions in source water, process water, and finished water locations. Some of the more uncommon online analyzers used at the case study utilities included those that measure phosphate, alkalinity, and, in a pilot-plant facility, UV_{254}.

In general, online analyzers were found to produce lower measures than bench instruments for all water quality parameters evaluated—turbidity, chlorine residual, phosphate, alkalinity, fluoride, and pH. For online turbidimeters, the location of the turbidimeter was not found to be important in distinguishing the comparative performance of bench and online measurements, even though this would have been expected given the issues with low-level turbidity measurement accuracy in finished waters. Instead, a significant degree of variability between bench and online turbidity measurements was found regardless of location (standard deviations of relative delta were nearly ±60%). For chlorine analyzers, no significant difference between online and bench measurements was found for all but one of the online instruments evaluated. The relative difference between online and bench measurements for chlorine analyzers was between 7% and 11%. For the other analyzers evaluated, phosphate and alkalinity analyzers showed less than ±5% variation in online and bench measurements, and pH and fluoride analyzers showed between ±15% and ±20% variability in differences between online and bench instruments.

Online measures can be used to evaluate the performance of online instruments. One example method of evaluating online instrument performance in a real-time environment is through the calculation of volatility—the running variance of online measures. The expected variability of a water quality parameter can provide norms of variance that such a measure should realize when measured using online instruments. When the volatility exceeds these norms, the excessively high or low variability of the measurement can serve as a trigger for online instrument support. The volatility results can also be used to characterize the quality of online measurements in terms of confidence in the measurement result.

CHAPTER 4
ONLINE DATA QUALITY, MANAGEMENT, AND APPLICATIONS

Online data collection poses a challenge to the staff responsible for managing this information. The quantity of information is far greater than other data collected. Frequencies of data collection can be a minute or less for some measures, resulting in thousands, even tens of thousands, of data points collected daily. If important operational decisions are keyed to these data streams, good decisions can result only when the quality of the online data is sufficient to meet the utility's needs. Using examples from the case study utilities that participated in the study, this chapter explores data quality, management, and applications.

ASSESSING ONLINE DATA QUALITY

When a utility uses water quality data to make decisions or report its performance, or both, the underlying assumption is that the data properly reflect the conditions the utility is trying to measure. To properly represent a water quality condition, the following issues must be assessed:

- Performance characteristics of instrument measurements
  - The "cleanliness" of online measurement streams in terms of meaningful data signals being collected and stored
  - The comparability of online and bench measurements when transitioning from bench measures to online measures
  - The expected variability of a water quality measure at the location of the online analyzer and whether the measurement variability by the instrument appears to be significantly different from that expected (as a result of background variability)
  - The need for, and ability to aggregate, online measures (i.e., calculating values such as averages, variance, and differences in measures) in a real-time environment
- "Lag" time needed before an online instrument can detect utility actions (operational changes)
  - The frequency (time step) of online data that are used virtually or in real time to support operational decision making
  - The frequency (time step) of online data that are recorded and used for operational decision making within a limited period of time (typically 24–48 hours)
  - The frequency (time step) of online data that are stored and used for offline (post hoc) data analysis

The issue of instrument performance characteristics is important to all applications of online monitoring data. Historically, many utilities have relied on bench instrument measurements to make decisions about plant operations. Today, online instruments are increasingly replacing these ways of taking measurements. In the United States, utilities exhibit mixed attitudes about the relative quality of bench and online measurements. Still, bench or laboratory techniques are used for the majority of compliance monitoring activities for water quality measures in the United States (although this is changing slowly). The degree to which the bench or laboratory measures are used for purposes perceived to have high data quality requirements indicates the perception that these measurements are more reliable than those produced by online monitors.

For drinking water utilities that still rely predominantly on bench measurements for important operational or regulatory activities, transitions to online monitoring results should be executed prudently and methodically. The performance characteristics of bench and online instruments are different, as shown in the previous chapter. Therefore, utilities must first collect comparable data to discern the differences expected between bench and online measurements, and then make adjustments in utility operations on the basis of such a priori information. In addition, based on the comparisons between bench and online measures, utilities may also need to modify the settings used for operational targets, the frequency at which operators are modifying treatment conditions, the maintenance support of the online instrument, or all three.

For online data used to inform operators in the control room or to set alarm conditions (in some cases), the quality of individual measures can be critical. The observation of a single measure outside of expected norms can set off an alarm that requires operator attention. The time-step requirement for these data may be very short and no aggregation may be required. Filtering of the data may still take place to address issues such as missing values, constant measurements that clearly do not reflect an operational sensor, and operating conditions where online measures are not expected to be valid (e.g., filter effluent turbidity during backwash events).

The offline use of data collected by online instruments is a critical component of most utility operations. Compliance reports, public information on water quality, evaluations of operational optimization success, and general inquiries into drinking water quality issues often rely on data collected by online sensors. However, storing every data signal sent by an online monitor is often not necessary to support such offline applications. Temporal aggregates are most commonly the data stored in offline databases for long-term uses. The utility needs to determine the appropriate time step for temporal aggregates for each water quality measurement. The population of data used to generate such aggregates is usually unique (e.g., hourly mean values or hourly standard deviations). The overlap with other time periods that occurs in running aggregates can be undesirable when statistical analysis of such data is anticipated. Validation of temporal aggregates uses population-based methods, although the data quality of single data measurements is important to the method of aggregation (e.g., inclusion or exclusion of data measures that have low confidence).

Making decisions about data quality requirements *before* the installation of online monitors is important, and such decisions must certainly be made before the instrument measures are incorporated into the SCADA or offline database environments. The process of defining such requirements is time consuming and detail oriented, but allowing these decisions to either be set to some default condition or to be determined by others outside of the utility business unit can lead to the collection, storage, and use of erroneous data and—ultimately—to poor decision making in utility operations.

Some essential steps that utilities should take to ensure the collection, storage, and application of high quality online measures include

- Data "cleaning." This is the process of eliminating online monitoring data that results from instrument signals during periods where the instrument is either nonoperational or is not receiving an appropriate water supply for measurement. For example, online turbidimeters often continue to send a measurement signal to a SCADA or data-logging system when a filter is being backwashed. These turbidity measures are not meaningful, and, in fact, do not represent the measurement of any water quality condition that is relevant to plant operations. Storing these data without a flag to denote the backwash condition could lead to inappropriate use of the data and incorrect operational decisions. Furthermore, in today's regulatory climate, these data could be used to demonstrate a false noncompliance condition. Therefore, proper cleaning of the online measurement data to produce relevant and appropriate measures in the data stream is an important first step toward the production of high-quality online data.

- Constant measurement checks. Online instrument measures should reflect at least the expected variability of the water quality parameter being measured. When a constant value is found over multiple measurement intervals, a flag should be raised about the performance of the instrument and the quality of the measurements.
- Instrument drift checks. By the nature of their environment, the sensors for online instruments are in contact with water on a continuous basis. The buildup of scale or deposits at the surface of the sensor can occur over time. In addition, those online instruments that have mechanical components (such as valves, reagent feed, and purging lines) can incur fatigue of these components, leading to a lack of consistent control. These and other effects on online instruments can lead to measurement "drift" where the signal generated by the sensor is inappropriately muted or amplified in a consistent direction. Several methods exist to identify instrument drift, including evaluation of running mean values over time and trending their deviation from established norms (set points would have to be established from a reliable period of record), and comparison of online measurements with bench measurements over an extended time period.
- Outlier determination. One of the more difficult data quality checks is the identification of outlier measurements. As discussed in the previous chapter, the use of a volatility measure may provide a tool for flagging potential outlier values. However, utilities should be cautious about the handling of such flags and associated data. Simply considering an observation to be an outlier does not mean that the value is automatically inaccurate; measurement abnormalities do exist. The issue for utilities will be to understand the relevance of these abnormalities to treatment decisions and plant performance characterization.

The generation of high-quality online monitoring data is a tremendous resource for drinking water utilities. Implementing steps to define the data quality produced by online sensors is a necessary component of a best practices program to support an online monitoring program.

WHEN ONLINE INSTRUMENTS ARE NEEDED

Utilities decide to use an online instrument for measuring water quality parameters based on a number of motivations, including identified informational needs, regulatory requirements, or operational support strategies. These uses of the online data affect all aspects of its collection, reporting, synthesis, and application in utility operations and management. Having a cogent and practical plan for online data use, then, is critical to a utility's successful application of online monitoring.

Conceptually, a utility's implementation and use of online monitoring can be divided into three activity levels—level 1, data acquisition and corrective actions; level 2, data handling and use; and level 3, utility management (Figure 4.1).

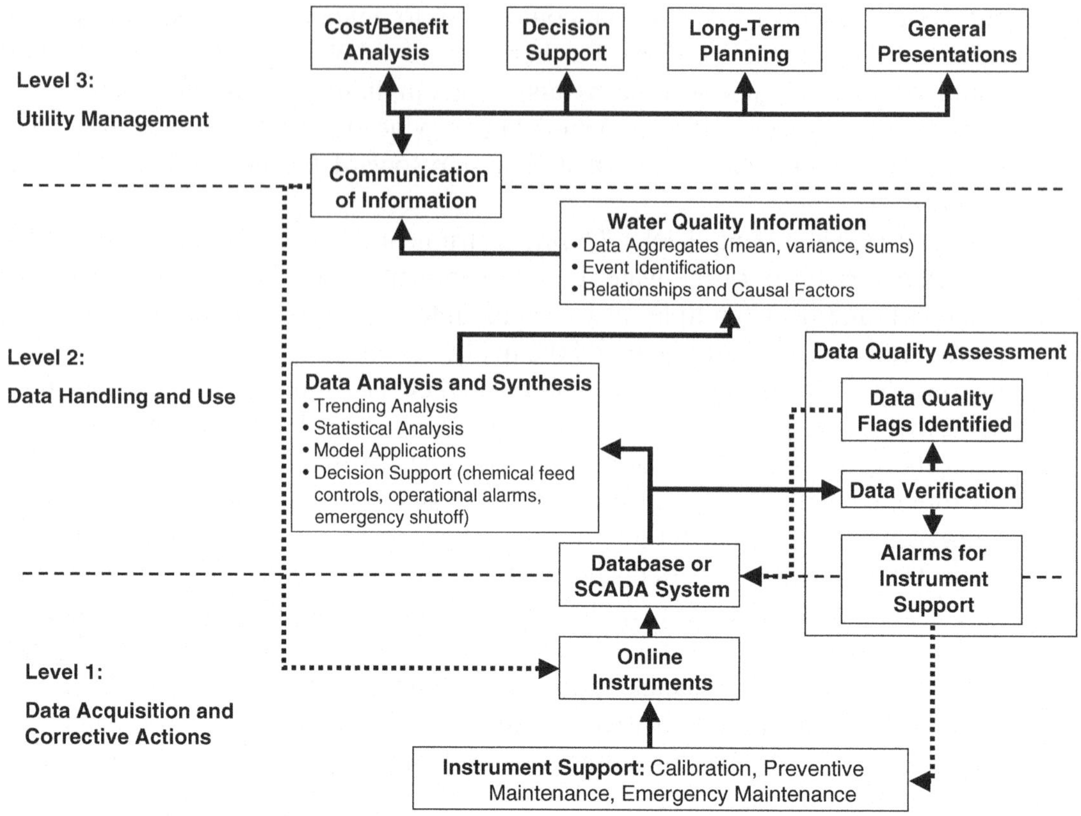

Figure 4.1 Conceptual framework for informational activities associated with online monitoring by drinking water utilities

1. Level 1 activities include the installation and operation of (a) actual online sensors/analyzers; (b) the communication interface for this data collection (such as a SCADA system or overall information management system); and (c) the physical systems necessary to operate the data collection system properly.
2. Level 2 activities involve a wide range of possible actions depending on the needs of the utility. However, the basic elements in this level include data quality assessments, data aggregation and synthesis, data transformation to information (or “data fusion”), and, finally, information communication.
3. Level 3 activities are supposed by the information communication activity in level 2. As with level 2 activities, the level 3 involve a broad scope of possible needs. In fact, these needs define the range of level 2 activities required for the online data collected in level 1.

The handling and use of online data must be managed so that the information effectively communicates effectively with each of the other levels of this operational framework. For example, installing online instruments and their interface with a SCADA system for the acquisition of data constrains the handling and use of the data collected based on the expected quality of the sensors/analyzers, the frequency with which reliable measures can be collected, and the capacity of the system for reporting and storing such data. In addition, the actual data use

Table 4.1
Comparison of SCADA and data logger capabilities

Function	SCADA	Data logger
Data storage	Moderate	Limited; must be downloaded regularly
Programmable functions	Yes. Real-time data quality determinations and calculation of aggregate values possible	No. However, the data can be downloaded into programmable software. No real-time data quality determinations possible
Automated system control	Yes	No
Generate standard reports	Yes	Not directly. Requires data download to programmable software
Generate standard data sets for export	Yes	No
Ad hoc query capability	Yes	No

defines the needs for the online sensors/analyzers and their measurement frequency, as well as providing the information resources necessary to flag corrective actions (e.g., maintenance and calibration) for sensor support. In addition, using the data to support management decisions or planning needs (level 3 activities) defines the objectives of the online monitoring activity and ultimately defines the communication needs from the application of data to the decision-making process of utility management.

CAPTURING ONLINE INSTRUMENT DATA

Online data can be captured in two principal ways, which differ significantly in their level of sophistication: in a SCADA system or in a simple data logger (Table 4.1). SCADA systems capture and store data, and, through programming capabilities associated with the system, permit the control of plant functions, generation of standard reports, aggregation of discrete online measures, and exportation of collected and aggregated data to external software environments. SCADA systems are usually supported by an underlying database system that may either be proprietary or based on commonly available data management systems. Typically, these systems are programmed to interface with readily available spreadsheet software programs to enhance external applications of the online monitoring data. Data logging systems do not have nearly as much flexibility or sophistication as a SCADA system; however, the typical array of SCADA functions is not always required. A data logger translates analog and digital signals into values. Data from data loggers are typically downloaded to an external data management system, which could be as simple as spreadsheet software or as sophisticated as highly evolved database software. Whether using a SCADA system or a data logger, exporting online monitoring data to other software environments for offline analysis is both feasible and necessary to support most utility requirements for online monitoring data applications.

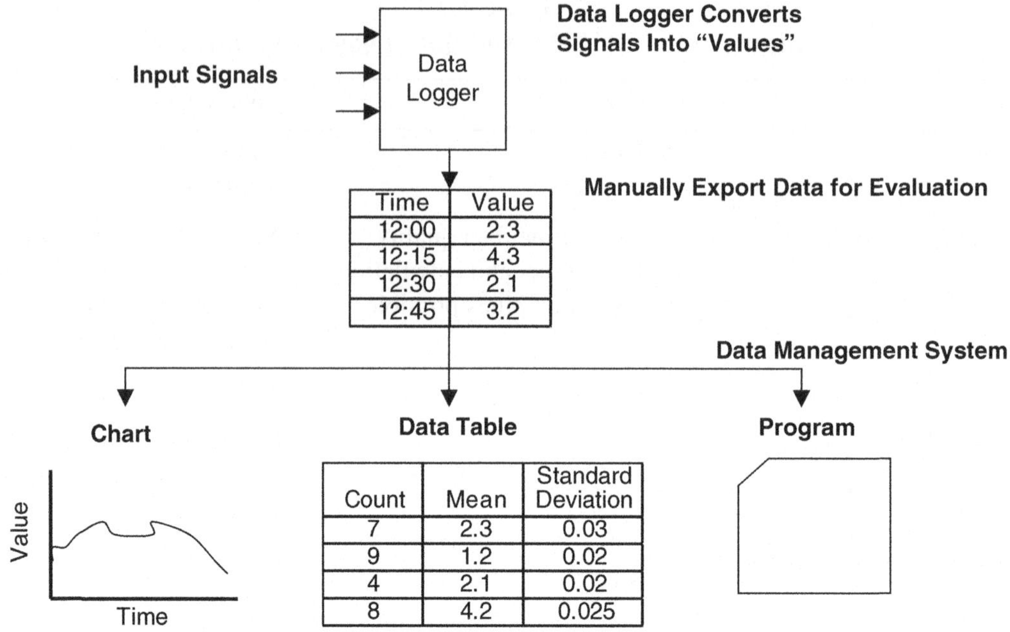

Figure 4.2 Schematic of data flow through a data logger

A powerful SCADA tool is the ability to program the system to use online data to evaluate instrument operation automatically and in real time. For example, running averages of the variance can be calculated and reported back on a SCADA screen. Values outside the defined tolerance can be flagged as questionable or used as alarm set points.

Data loggers cannot perform data quality checks in a real-time operational mode; these functions have to be performed external to the unit. Although this requires additional attention and time, it is still an effective tool. Figures 4.2 and 4.3 schematically depict the general structure and functions of SCADA and data logger systems.

USING ONLINE DATA SCREENS FOR SCADA SYSTEMS

A great advantage of a SCADA system is the ability to use the online data to establish instrument operation goals, and then use the goals as set points for operational flags or alarms. This information can be reported back on the SCADA system in the same fashion as other conventional signals. An example scenario, including an illustration of a SCADA screen, is described in the paragraphs that follow.

This example assumes that there are two residual analyzers located in a clearwell. Historically, maintaining a stable chlorine residual at this plant has been a challenge, and by installing two chlorine residual analyzers, the reliability of the overall system has been improved. The left-hand side of the SCADA screen (Figure 4.4) is reporting the volatility (running variance) of the signals. Historic data were used to determine a low set point of 0.02 and a high set point of 0.06. The right-hand side of the screen is reporting the delta of the two measurements.

On the left, instrument CL2CW-1 is operating within the set points; however, in this example, CL2CW-2 has exceeded the high-level set point. Although individual excursions may not warrant action, if this instrument continued to operate over the high-volatility set point, this

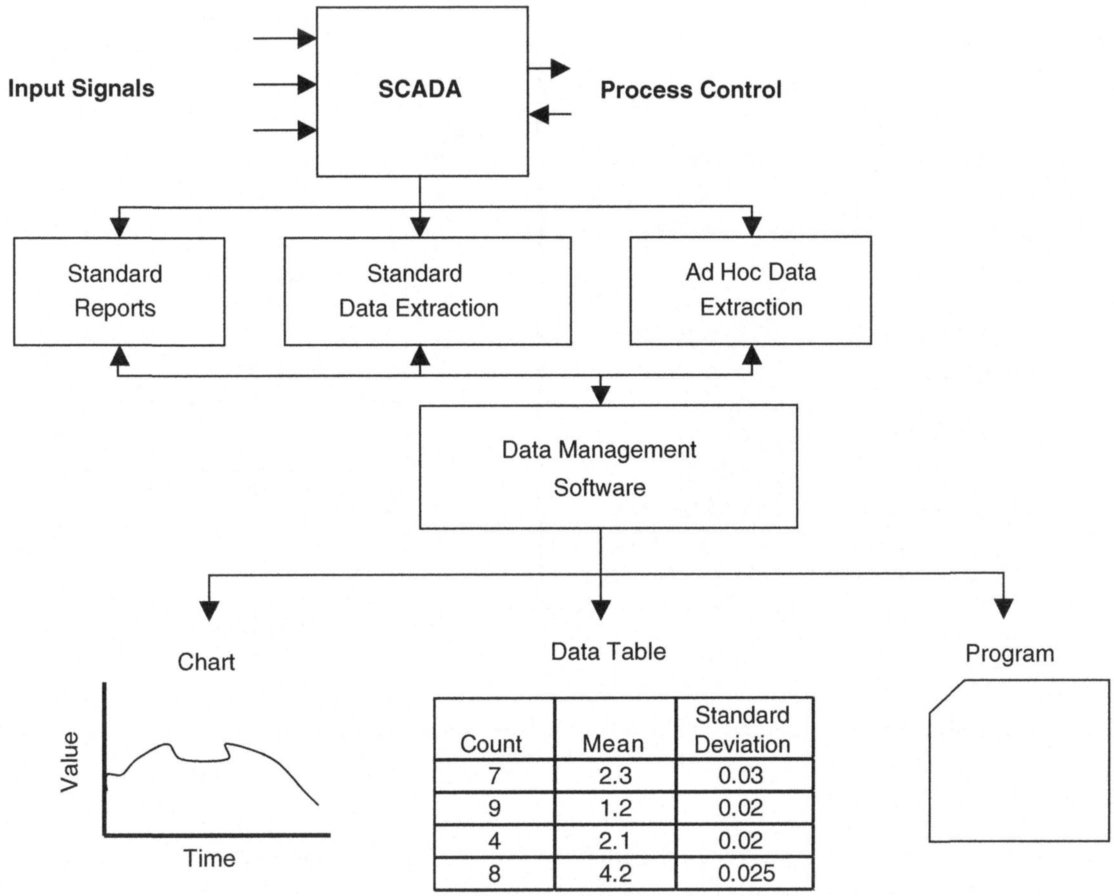

Figure 4.3 Schematic of data flow through a SCADA system

could indicate that the instrument's precision has been affected and should be addressed. The calculation of the deltas, on the right side of the figure, shows that the instruments are operating within a prescribed ±0.2 difference.

The use of SCADA systems as tools for checking the operation of online instruments is not widespread. As greater emphasis is placed on SCADA systems, it would be prudent for systems to include internal instrumentation verification.

APPLICATIONS OF ONLINE MONITORING DATA FOR INSTRUMENT SUPPORT

Primarily, online data is used to provide real-time or historic assessments of plant performance. For real-time operation, set points are defined and alarms or warnings are triggered when these set points are exceeded. In addition, online data are often used to generate reports or to provide historical perspective on operations.

Although these are important applications of online data that deal specifically with evaluating the operation of a water treatment plant, another class of application is also very powerful. Applying online data to evaluate the performance of online instruments is a valuable tool that is often overlooked in today's facilities. Many of the techniques for applying online data to evaluate online instrument performance have been introduced previously in this report. This

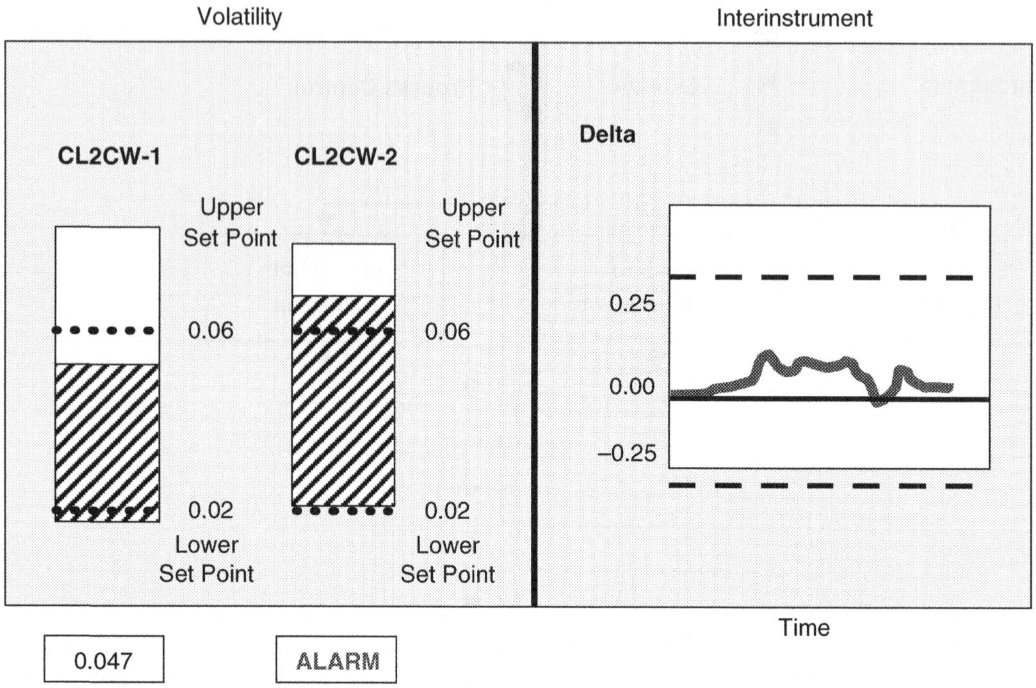

Figure 4.4 Illustration of SCADA screen for online data verification

section will take these concepts a step further by presenting detailed examples of how to apply online data for various instrument-check applications.

Volatility Trigger

Volatility is a measure of instrument precision and has been mentioned numerous times already in this report. The description of the sample SCADA system screen in the previous section used volatility as an example parameter and illustrated how it could be used on a real-time basis. However, volatility is not limited to real-time applications. The volatility of historic data can be assessed using a spreadsheet program. Implementing the use of volatility is fairly straightforward. The recommended steps for developing volatility triggers follow, in the context of an example using case study data.

Example

Step 1. Assemble a Large Set of Online Data. Quality check the data for anomalies, as discussed earlier in this chapter. In this example (see Table 4.2), 856 online chlorine residual values were aggregated. The running variance was calculated for 30 values at a time, resulting in 826 running variance data points.

Step 2. Calculate the Median, Mean, and Standard Deviation of the Variance Data. Table 4.2 shows the results of evaluating chlorine residual data from case study B.

Table 4.2
Evaluation of variance for case study B chlorine residual

Parameter	Value
N*	856
Ncalc†	30
Median	0.0187
Mean	0.0201
Standard deviation	0.0098
Maximum	0.068
Minimum	0.0077
Frequency of collection	6 per day

*N = number/count.
†Ncalc = The number of online values used for each variance calculation.

Step 3. Select High and Low Set Points. For example, initially set points may be ±1 standard deviation, but 1 standard deviation may be found to be too narrow or too broad, meaning that the set points will have to be modified. Experience will be the final guide for selecting set points.

In a SCADA system, this calculation can be performed continuously and triggers can be programmed in; for data logger systems, the data evaluation must be performed regularly in an external software program.

ONLINE VERSUS BENCH TRIGGER

Comparisons of online versus bench data are the most common instrument verification method applied. Of the 392 treatment plants that responded to the national survey, the percent of instruments for which verifications are performed at least quarterly ranged from 20% to 71%, depending on instrument type. This process works only if the bench instrument is properly maintained and calibrated. In addition, samples from the online instrument must be carefully collected and should be measured on the bench instrument immediately. Some utilities use bench comparisons as triggers to indicate when an online instrument should be calibrated; others use bench comparisons only following a calibration to verify the results of the calibration. Comparison data can be used to evaluate and establish an application-specific calibration frequency. This is an important step in developing an SOP, and a detailed example is given in chapter 5.

Table 4.3
Instruments that can be used for interinstrument verification

Location	Instrument type
Split trains	Sedimentation basin turbidity
Filtration	Turbidity
Filtration	Particle count*

*To use different particle counters to check one another, a study to evaluate the interinstrument variability would be necessary.

Interinstrument Verification Tools

There are three primary tools for checking online instruments using other online instruments:

1. Comparison using two of the same type of online instruments
2. Comparison using different types of online instruments measuring related parameters
3. Comparison using predicted values to check online values

Using the same online instrument to evaluate a single parameter can be accomplished in two ways:

1. Measuring at like but physically different locations (for example, sedimentation basin effluent from multiple trains), or filter effluent
2. Measuring the same parameter at the same location utilizing two instruments

Many treatment plants have split-train configurations or locations, such as filter effluent, where there are numerous instruments measuring parameters that should be relatively close all the time. The measurements from these instruments can be compared to provide an evaluation of instrument performance. Table 4.3 shows examples of common locations and instrument types that can be used in this manner.

Although it is fairly uncommon for plants to install two like instruments at a single location, this practice is occasionally implemented. For very important applications—for example, finished water chlorine residual—two instruments can be installed side by side to improve reliability. One instrument can be used to check the other, and alarms can be based on the difference, or, in some cases, instrument signals may be averaged. Chlorine dose is often controlled by the online chlorine residual measure. Redundant instruments ensure uninterrupted disinfectant feed if one signal is lost.

Using different types of online instruments to evaluate one another is another method of instrument check. One of the most common examples of this is particle counters and turbidimeters. These two types of instruments are often used simultaneously to measure the effectiveness of filtration, and they can also be used to assess one another. Particle counters are more sensitive than turbidimeters, so changes in turbidity should result in significant changes in particles.

Data collected by a SCADA system can be used to calculate predicted values, and then the predicted values can be continuously compared to actual values. For example, based on a known water quality, an algorithm can be developed to predict the change in pH that should occur based on coagulant dose. Assuming that the SCADA system is accurately measuring chemical dose, the predicted pH value can be calculated and compared to the actual pH value downstream (adjusted for appropriate time lag).

METADATA ASSOCIATED WITH ONLINE MONITORING PROGRAMS

In addition to the large quantities of measurement data that must be managed from online instruments, there are also a number of other sources of important information associated with the support of online instruments. The term "metadata" refers to the array of information that further defines, qualifies, and characterizes measurement data. Two predominant areas of metadata collection, storage, and management are associated with online monitoring systems: instrument-related and measurement-related. Examples of each of these types of metadata and the components that should be collected and managed by drinking water utilities are discussed in the next section.

METADATA RELATED TO ONLINE INSTRUMENTS

The application of online instruments for water quality monitoring entails a substantial body of information related to the installation, operation, maintenance, and location of each instrument. To enhance a utility's online monitoring program, the sources of information described here should be organized and maintained in such a way to permit easy access and reference to staff members responsible for online instrument support.

- Measurement method. The actual analytical method that the online instrument uses to measure the water quality parameter of interest should be described.
- Instrument specifications. When a utility purchases an online instrument, the manufacturer supplies documentation on the instrument's specifications and the requirements for its installation, maintenance, and calibration. These documents should be stored in a central library location. In some cases, these documents are available in an electronic format for storage and management.
- SOPs. Each type of online instrument in use at a utility should be accompanied by an SOP that fully describes the utility's program for maintenance, calibration, and operation of the units. In the next chapter, the components of SOPs for online instruments are described.
- Maintenance tracking. Proper operation of online instruments requires routine preventive maintenance as well as emergency maintenance of the devices. The manufacturers often provide information on recommended preventive maintenance program requirements. Utilities should develop a tracking system to ensure that the expected preventive maintenance activities are carried out within the recommended schedule and to furnish additional information to instrument support personnel about the operational performance of the instruments. By tracking this information, problematic instruments can be identified and measures taken to either improve their performance or replace them.

- Spare parts inventory. Onsite maintenance of online instruments is required, and having spare parts on hand is a critical component of a successful maintenance program. Although some repairs cannot be made in the field, the spare parts for those repairs that can be made should be identified and an inventory tracking system developed to ensure that they are available on demand.
- Instrument location. When online instruments are connected via a SCADA system, their signal is given a “tag” identification. This identification is typically included in the SCADA system dictionary of instrument locations. Because these locations are often coded, some users may not have a clear definition of the actual physical location of the instrument. Such descriptions of physical location should be included in the information stored and managed by the utility, whether a SCADA system or a data logging system is used to collect online monitoring data.

Online Measurement Metadata

The actual measurement value generated by online instruments requires further definition and characterization. In particular, the metadata that should be stored with online measurement data include:

- Measurement location. The measurement location is as important as the physical location of the instrument. The instrument location may be defined as, for example, the “south wall of filter gallery for filter 2 effluent.” However, the practical use of this description to identify measurement locations is minimal. Instead, utilities will typically identify a code, such as “FE2,” to identify the measurement location. Standardized codes for measurement locations are necessary for aggregating online monitoring data in various ways (such as for comparing instruments and calculating mass balances) and for synthesizing online and bench measurements in a meaningful manner.
- Calibration date. The date or time from the last calibration event is useful in assessing the likely accuracy of the online instrument measure.
- Measurement units. The units in which the online instrument measurement are represented in the SCADA or data logging system must accompany all online measures.
- Field checks. Observations from operators or online instrument technicians who find anomalies when checking the instrument in the field are important qualitative data that speak to the likely performance of the instrument and the related quality of the instrument measures.
- Data quality flags. Using the techniques described earlier in this chapter, various real-time assessments of online instruments to measure data quality can be generated. The compilation of these assessments into a data quality flag is useful for characterizing the quality of the measures.

Maintaining appropriate metadata associated with instrument measurements enables utilities to use their online monitoring data with added confidence and value. Without such metadata, however, online instrument measurements are subject to important uncertainties that may not be able to be resolved after the fact.

Managing Online Instrument Metadata

This information is typically stored in various locations (for example, control rooms and maintenance departments), as well as in different formats (paper versus electronic). Today, all of these metadata can be acquired, developed, or stored in electronic format. “Paperless” environments have a number of advantages over paper-only or paper and electronic environments:

- Physical storage requirements are considerably less and backups can be automated.
- Information can be stored on a network so that it is readily accessible to everyone.
- Information can be linked together to create comprehensive documents where all the information for an instrument can be accessed.
- More sophisticated applications can be designed and implemented. For example, if a set point is triggered in the SCADA system, a work order can be generated and tracked.

Converting to a paperless environment may not be necessary, or even appropriate, for all plants. However, it is recommended that where possible, electronic information management be implemented, as this is where the future of information management is clearly heading.

ARCHIVAL REQUIREMENTS

Systems for backing up and archiving data are important components of overall data management. The primary driver behind archival requirements is legislation, which typically requires that compliance data be stored for 5 to 7 years. Many viable methods exist for backing up data, the most common of which is probably automated tape backups. Data backups should not be confused with long-term storage or archiving. Systems for archiving vary, but the common steps include selecting an archival interval and selecting a storage method. Data can be stored on individual disks or tapes or stored on a local intranet or the Internet. These services are also available commercially. Document management firms will download, index, and maintain a document warehouse that can be accessed via a local area network, intranet, or the Internet.

CHAPTER 5
SOP DEVELOPMENT

Using online instruments to measure water quality parameters at drinking water facilities results in a rapid stream of information on plant operations and performance. Making the best use of this information necessitates that drinking water utilities develop and maintain good SOPs. This chapter presents guidelines for developing practical operating procedures for online instruments.

An online instrument SOP serves as a reference for operators as well as a training document and a guide for online instrument support. A well-developed SOP can give utilities a road map that ensures consistent instrument operation, calibration, maintenance, and, subsequently, better overall quality of online measures. SOPs should be developed for each instrument type with specific requirements based on the instrument's location within the plant and the ultimate use of the online data. For example, online instruments that measure raw water quality may require a different routine maintenance schedule because of the need for more frequent cleaning of sample lines, instrument chambers, or sensor surfaces. In addition, the intended end use of online monitoring data determines the reliability and confidence required for the data. For example, online monitoring data used to trend water quality conditions or treatment performance may not require the same degree of accuracy and precision as online monitoring applications used to control chemical feed systems. To this end, the SOP may contain variable calibration requirements for the same instrument type, depending on the application of the online monitoring data for individual instruments.

There are common features and components for online instrument SOPs. At a minimum, SOPs should include

- Introduction. This section should feature a description of the parameter being measured and its importance to water treatment, along with some general information about how the parameter is measured. This information serves as background for training new operators on the importance of the measurement to the overall goal of treating water effectively.
- Operation. Operation information is important for the day-to-day upkeep of the instrument and appropriate application within a utility's operational requirements. One of the most important parts of online instrument operation is routine instrument observation. Others are the data acquisition, quality assessment, handling, analysis, and management for online instruments.
- Calibration. Calibration procedures and the establishment of calibration frequencies are critical aspects of instrument operation. Online instrument calibration establishes the expected accuracy of online instrument measurements. Inappropriate procedures or frequencies can lead to the collection and use of erroneous data, resulting in poor operational control and decisions for water treatment plant operations. Identifying appropriate calibration procedures and frequencies is discussed further in the next section (see Table 5.1).
- Maintenance. In general, two types of maintenance activities should be included in online instrument SOPs: (1) preventive maintenance, which is prescriptive, having defined activities according to a specified schedule; and (2) emergency maintenance,

Table 5.1
Example data collection and analysis for calibration frequency determination

Percentile	All data (12 months)	<9 months	<6 months	<3 months	<2 months	<1 month
0.05	0.05	0.04	0.04	0.04	0.02	0.02
0.1	0.09	0.07	0.06	0.06	0.05	0.04
0.2	0.19	0.16	0.14	0.12	0.09	0.07
0.3	0.32	0.26	0.21	0.18	0.18	0.14
0.4	0.49	0.41	0.34	0.25	0.21	0.19
0.5	0.74	0.56	0.44	0.40	0.36	0.32
0.6	0.98	0.76	0.63	0.54	0.45	0.41
0.7	1.25	0.96	0.79	0.74	0.58	0.56
0.8	1.50	1.22	1.12	1.07	0.79	0.75
0.9	1.89	1.54	1.23	1.17	0.94	0.88
0.95	2.40	2.09	1.92	1.45	1.30	1.12
0.99	3.95	3.54	3.38	3.31	3.00	1.35

which is responsive and is carried out on an "as needed" basis. An emergency maintenance program's requirements may be difficult to describe within an SOP until experience with specific instruments has been gained over time. When a utility's experience with the maintenance of an online instrument is limited, the initial SOP should incorporate the manufacturer's recommendations for routine maintenance. Over time, updates and improvements in SOPs should reflect the maintenance issues and schedules that best suit the instrument's operation.

As described above, an SOP is expected to evolve over time as utility staff members gain experience and more profound knowledge of the operation, maintenance, calibration, and use of online monitoring information. As a work in progress, the SOPs developed by utilities represent the user's state of the art with respect to online instrument support. These documents also form a framework from which online instrument support can be standardized and managed within a utility's organizational structure. This chapter uses an example template of an SOP for online turbidimeter applications to serve as a guideline to utilities for developing online instrument SOPs. Checklist development, data collection and analysis activities, and example descriptions for each major component of the SOP are covered in this discussion and amplified by the example given in appendix C.

Beyond SOPs, there are other informational resources important to the operation and application of online monitoring in systems. Many of these, such as installation information and spare parts inventory, were described in the previous chapter under the metadata analysis for online instruments. This additional information can often be found in manufacturers' manuals and may be more suitably stored and managed elsewhere within a utility's information management system. The purpose of an SOP is not to repeat information accessible elsewhere but to make available a document specific to the upkeep and operation of online instruments.

OPERATIONAL PRACTICES

Within the operational practices section of an SOP, three important areas of online instrument support should be presented:

1. Routine visual checks of online instruments. Each online instrument should be checked visually according to a routine schedule. Visual checks include physical observations of the instrument to ensure its proper function, and should include general activities such as observing and regularly measuring flow rates through the instrument; inspecting and cleaning sample lines; checking reagent levels and dosing systems (pumps, valves, and tubing) operations; testing system alarm functions; and checking visual displays of instrument measures. More specific checks are typically identified through experience (for example, probes that need cleaning as a result of metal deposition). As specific items are identified, they should be added to the operational procedures within the SOP.
2. Data quality assessment of online instrument measurements. Several techniques were described in the previous chapter for using online monitoring data to assess the performance of online instruments. Deviation in data quality measures from established norms can be used to trigger an emergency maintenance activity. These data quality measures can also be used to inform plant operators about reliance on the online measurements for treatment plant operational decisions.
3. Online data application assessments for plant operations. The expected use or application of online monitoring data within a plant's operational function or decisions, or both, should be clearly described in the SOP. In addition, contingency procedures should be described to inform utility operational staff of the actions necessary when online monitoring data fail to meet performance norms. These should include communication requirements, work order procedures for triggered maintenance support, and any procedures to initiate special investigative activities deemed appropriate by operations staff.

SETTING CALIBRATION FREQUENCIES

Calibration frequencies will vary from plant to plant and will depend primarily on instrument application, available support, and instrument operational goals. Establishing a calibration protocol (methodology and frequency) requires three steps:

1. Identify reasonable instrument operation goals. An operation goal is defined by two parameters. First, the instrument must achieve a certain "agreement" or accuracy as compared with the bench instrument. Second, the instrument must achieve this delta a certain percentage of the time. For example, a goal for a finished water turbidimeter might be a bench versus online instrument delta of 0.05 ntu 95% of the time.

Considerations in selecting a goal include a check of the operational limitations specified by the manufacturer and the specific instrument application.

2. Conduct data collection and analysis. The online instrument should be calibrated and bench versus online data must be collected regularly to establish performance based on the existing operation protocol. These data are then analyzed to identify the calibration frequency at which operational goals are met (see Table 5.1).
3. Formalize and document protocol or establish new goals. If the calibration frequency identified is acceptable, a formal protocol should be developed and training on the protocol should be made available. If it is not acceptable, a different goal and frequency that meets the utility's needs should be selected based on the data collected.

It is not necessary to repeat this data collection effort for each individual instrument; however, tailoring calibration support to the instrument location, measurement type (i.e., online instruments using different operating principles or analytical methods should have unique calibration protocols), and application of the online monitoring data (e.g., compliance monitoring versus trending of water quality measures) are necessary. Table 5.2 shows a sample summary table describing different protocols that a utility could include.

Table 5.1 presents an example procedure that utilities could use in developing their own calibration protocols for online instruments. The concept is that calibration frequencies can be determined by evaluating the likelihood that differences between online and bench measurements between calibration cycles would indicate the likely timeframe before such differences exceeded established norms. Likewise, investigation of online instrument drift and volatility over time could inform the decision about calibration frequency. Although initial SOP development should incorporate recommendations from online instrument manufacturers, utilities should proactively develop a procedure to determine calibration frequencies to suit the QA requirements of the utility's online monitoring program.

Table 5.2
Summary of data collection for calibration determination

Instrument location	Summary	Target accuracy
Raw	Instrument application is trending. Monthly bench comparison—calibrate when difference is ≥1.5 ntu or every 3 months, whichever occurs first.	<1.0 ntu difference 95% of the time
Filter effluent or finished	Instrument application is compliance. Weekly bench comparison—calibrate when difference is ≥0.03 ntu or every 3 months, whichever occurs first.	<0.03 ntu difference 90% of the time

SETTING PREVENTIVE MAINTENANCE SCHEDULES

Interviews with the case study utilities and assessment of available information from various online instrument manufacturers, as well as the experience of the authors, all confirm that preventive maintenance schedules for online instruments should be determined on a site-by-site and case-by-case basis. It is once again important to consider the application of the specific device, learn about the device over time, and develop a "personalized" or instrument-specific preventive maintenance schedule. The following three steps are suggested for developing such a schedule:

1. Review the manufacturer recommendations for maintenance and develop a schedule based on the information provided.
2. Begin a log for each instrument type/location combination. In the log include details of all actions associated with the instrument. Include comments about unusual or seasonal water quality or treatment impacts that may be related to the logged event.
3. Use the manufacturer information and the plant information contained in the logbook to develop a preventive maintenance schedule. Again, this should be considered a working document and updated frequently.

CONCLUSIONS

Every utility should develop SOPs for the online instruments used at its facilities. The example template in appendix C illustrates the components and types of information that any SOP should contain, but the specifics must be developed on a site-by-site, instrument-by-instrument basis. On the part of drinking water utilities, this represents a substantial effort in terms of investing in the proper operation, maintenance, calibration, and use of online monitoring data. However, without such a resource, online monitoring systems will remain questionable systems that produce values in which utility decision makers have little confidence. The value of online instruments lies in their ability to better inform utility operators, allowing them to make decisions in a more real-time manner. Without appropriate SOPs that are actually used to support online instruments, utilities will be unable to fully realize the value of online instrument applications.

APPENDIX A
NATIONAL SURVEY OF ONLINE MONITORING PRACTICES

INTRODUCTION

The National Survey of Online Monitoring Practices was mailed in March 1998 to all the utilities represented in the Water:\Stats database. In all, 264 utilities, representing 392 treatment plants, responded to the survey, which was used to benchmark the use of online instruments in the water industry. This appendix contains a copy of the survey.

AwwaRF Project 2516
Practical Application of Online Monitoring
Utility Survey of Online Instrumentation

Utility Name ____________
PWSID ____________
City/State ____________

Please estimate the time required to complete the survey:
_____ Hrs.

		INSTRUMENT INFORMATION					
				Operating Range for Instrument Measurements			
				Lower		Upper	
Row Identification	Plant Name	Instrument Type	Instrument Location	Value	Units	Value	Units
Example 1	EXAMPLE	Turbidimeter	Filter Effluent	0.01	ntu	1	ntu
Example 2	EXAMPLE	Turbidimeter	Raw Water	1	ntu	100	ntu
1							
2							
3							
4							
5							
6							
7							
8							
9							
10							
11							
12							
13							
14							
15							

INSTRUMENT INFORMATION					
Instrument Measurement Applications in Plant Operation					
SCADA Loop Controls	Chemical Feed Rates—Manual	Performance Trending	Compliance Reporting	Other Instrument Check	Other
		X	X		
		X			X

PLANT PRACTICES												
Do You Have SOPs? No = No SOP, Type=(U)tility or (M)anuf., O=Operation, C=Calibration, M=Maintenance					Approximate Frequency of Calibration per Year	Calibration Method			Routine Verification Checks With Bench-Top Instruments		Hours per Month Spent on Instrument Support?
No	Type	O	C	M		1 Point	2 Points	Other	Y=Yes N=No	Frequency (monthly)	
	U	X	X	X	4		X		Y	30	2
	U	X		X	1		X		Y	1	0.5

APPENDIX B
CASE STUDY SUMMARIES

CASE STUDY A

INTRODUCTION

Case study A's (CS-A) water treatment plant encompasses several large and small, direct and conventional filtration plants. Because CS-A is providing historical turbidity information for the project, this case study summary is not as detailed as those supplied by the other utilities.

High-quality, low-turbidity source waters are treated at the plants. Table B.1 summarizes 18 months of data submitted as part of the Information Collection Rule (ICR) data collection effort for three of the plants.

Table B.1
CS-A's 1998 raw water quality

Parameter	Average	Minimum	Maximum
	Plant A		
Turbidity (ntu)	4.77	0.34	29
Alkalinity (mg/L)	96.8	70	120
pH (SU*)	7.68	7.29	8.18
Temperature (°C)	16.1	10.1	22.2
TOC† (mg/L)	4.39	3.65	4.95
	Plant B		
Turbidity (ntu)	0.86	0.25	1.9
Alkalinity (mg/L)	15.3	10	21
pH (SU)	7.26	6.41	8.2
Temperature (°C)	14.6	9.9	19.1
TOC (mg/L)	1.15	0.85	1.5
	Plant C		
Turbidity (ntu)	5.82	0.83	42
Alkalinity (mg/L)	77.3	59	110
pH (SU)	7.51	7.08	7.85
Temperature (°C)	14.9	10.0	21.0
TOC (mg/L)	3.83	3.3	4.65

*SU = standard units.
†TOC = total organic carbon.

INSTRUMENTATION

Table B.2 summarizes CS-A's online instrumentation by location. Currently, there is no online instrumentation in the distribution system.

Figure B.1 illustrates the plant's supervisory structure.

Table B.2
CS-A's online instrumentation

Location	Instrumentation
Raw	pH
Rapid mix	Chlorine, pH
Ozone	Ozone residual
Filter effluent	Turbidity, particle count, pH
Finished	Chlorine residual, ammonia, fluoride

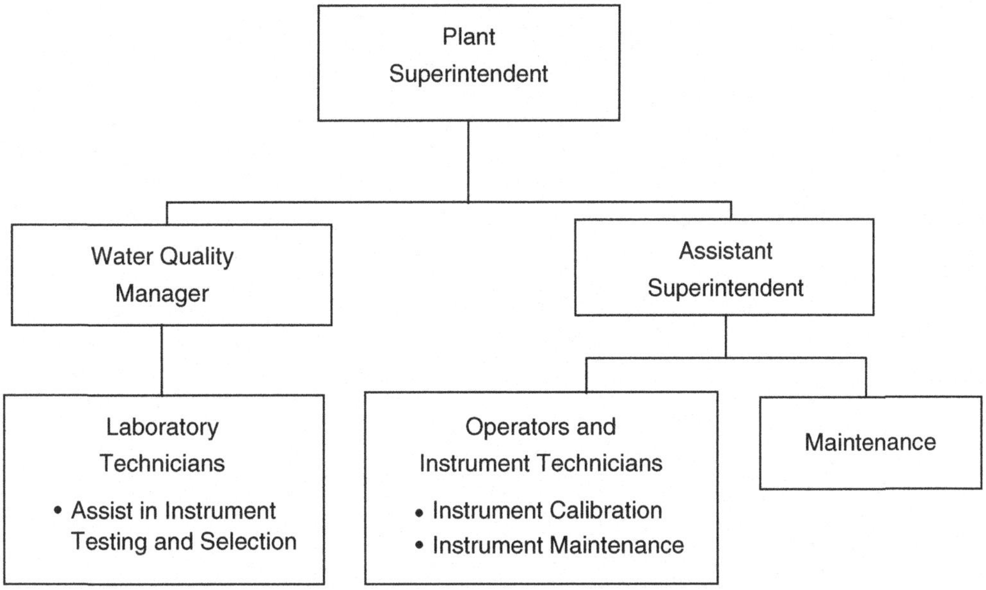

Figure B.1 CS-A's organizational structure as it relates to online instrument support

STUDY PARTICIPATION

CS-A provided historical turbidity data from comparison samples that had been collected (see Table B.3).

Table B.3
CS-A's participation

Instrument	Location	Number	Frequency of bench verification
Turbidity	Influent	1	1 per week
	Filter effluent	10	1 per week

CASE STUDY B

INTRODUCTION

Case study B (CS-B) is a direct filtration plant with a capacity of 70 mgd. Surface water is the primary supply; however, the city also has several well fields. From a water quality perspective, groundwater is important to the system. During spring runoff, the higher alkalinity present in the groundwater is used to augment the lower alkalinity surface water. Table B.4 displays the source water quality in 1998.

INSTRUMENTATION

Table B.5 summarizes all of CS-B's analytical online instrumentation by location, and Table B.6 describes the routine instrument support provided by maintenance and operations staff.

Table B.4
CS-B's 1998 raw water quality

Parameter	Average	Minimum	Maximum
Turbidity (ntu)	3.4	0.76	8.48
Alkalinity (mg/L)	82	68	118
pH (SU)	7.93	7.56	8.30
Temperature (°C)	12.0	4.0	20.0
TOC (mg/L)	3.0	2.2	4.0

Table B.5
CS-B's online instrumentation

Location	Instrumentation
Raw	Turbidity, particle count, streaming current detector (SCD)
Filter effluent	Turbidity
Backwash	Turbidity
Clearwell	Chlorine residual, particle count
Distribution system influent	Chlorine residual, turbidity

Table B.6
CS-B's instrument support

Instrument	Location	Support
Turbidity	Raw, filter effluent	Bench checks every 2 hours
Particle count	Raw, clearwell	Clean screen Clean sensor cell Check effluent flow Clean exterior
Chlorine residual	Clearwell, influent to distribution	Bench checks every 2 hours

Table B.7
CS-B's instrument uses

Location	Online instrument	Function
Raw	Turbidity	Informational
	Particle count	Informational
Filter effluent	Turbidity	Compliance reporting
	Chlorine	Manual chemical dosing
Backwash	Turbidity	End backwash
Clearwell	Chlorine	Manual chemical dosing
	Particle count	Informational
Entry to distribution	Chlorine	Manual chemical dosing

Treatment Challenges

Source water comes from both surface water (three different surface water basins) and groundwater. The variability in these water sources presents this plant with its greatest treatment challenges. High-turbidity events (spring runoff) plugs screens on raw water analyzers, requiring screens to be flushed more often and the system to be cleaned more frequently. Alternate sources also require potassium permanganate for taste and odor control. Application of potassium permanganate also requires equipment to be cleaned more often to remove manganese deposits, along with a greater frequency of zeroing the SCD.

Instrument Use

CS-B uses online instrumentation primarily for informational and manual dosing purposes (see Table B.7).

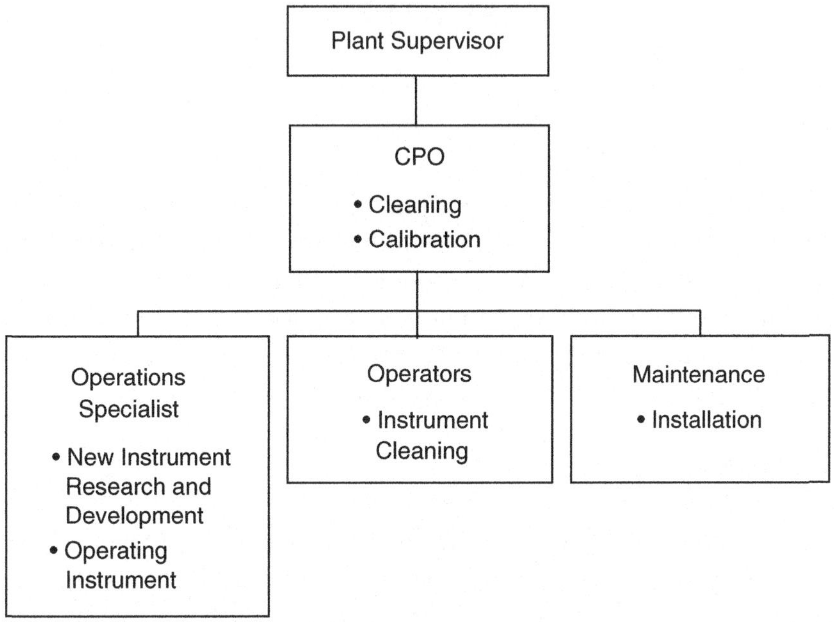

Figure B.2 CS-B's organizational structure as it relates to online instrument support

Table B.8
CS-B's participation

Instrument	Location	No.	Frequency of bench verification
Chlorine residual	Clearwell	3	Every 4 hours
Turbidity	Raw	1	Every 4 hours
	Filter effluent	1	Every 4 hours
	Clearwell	1	Every 4 hours

MANAGEMENT

Figure B.2 shows the management structure at CS-B as it relates to instrument support. The chief plant operator (CPO) calibrates the instruments, and both the CPO and operators perform the cleaning. In addition, this facility has a special position—operations specialist—that handles new instrument research and development of operating procedures.

STUDY PARTICIPATION

CS-B evaluated turbidimeters and chlorine residual analyzers at multiple locations. Table B.8 summarizes the plant's participation by type, location, and number of instruments, as well as by frequency of bench verification. CS-B collected data with the greatest frequency of all the case study utilities. In addition, CS-B has a fairly long calibration cycle (the time between calibration events), and thus the data may provide one of the best snapshots of instrument variability over time.

CASE STUDY C

INTRODUCTION

Case study C's (CS-C) water treatment plant is a conventional lime-softening plant. This city's primary raw water source is a river, but wells are also used to supplement flow, to add carbon dioxide for re-carbonation and, during cool months, to improve the softening process at low temperatures. This facility's processes include ozonation to provide disinfection contact time (CT) and granular activated carbon–capped filters for precursor material removal. Table B.9 shows the range of raw water quality at CS-C, recorded in 1998.

INSTRUMENTATION

Table B.10 summarizes CS-C's online instrumentation by location. There is no online instrumentation currently in the distribution system. Table B.11 shows the current instrument support practices.

Table B.9
CS-C's 1998 raw water quality

Parameter	Average	Minimum	Maximum
Turbidity (ntu)	2.6	0.7	5.5
Alkalinity (mg/L)	208	195	215
pH (SU)	8.1	7.9	8.2
Temperature (°C)	16.1	6.8	23.5
TOC (mg/L)	6.0	4.0	7.7

Table B.10
CS-C's online instrumentation

Location	Instrumentation
Raw water	Turbidity
Postrecarbonation	pH
Postozone	Ozone residual
Filter effluent (individual)	Turbidity, particle count (2 filters)
Clearwell	pH, turbidity, chlorine residual
Backwash	Turbidity
Finished water reservoir	Chlorine residual, turbidity
Pilot-plant effluent	UV_{254}*

*Ultraviolet at a wavelength of 254 nanometers.

Table B.11
CS-C's instrument support

Instrument	Location	Support
Ozone residual	Ozone contactors	Calibrate on an as-needed basis; typically the result of having to change membrane
pH	Recarbonation	None
	Clearwell	Clean every 6 months
UV_{254}	Pilot plant	None
Turbidity	Filter effluent	Monthly calibration with formazin standard or polymer-based standards

Treatment Challenges

Treatment challenges from an instrumentation standpoint evolve from the softening process and the inherent scaling effects associated with softening. pH probes resistant to the effects of calcium carbonate precipitation were difficult to find; however, during recent upgrades, new pH probes were installed and have seemed to operate satisfactorily. These new probes were evaluated as part of this project.

Instrument Use

Online instruments serve a number of purposes for CS-C. Functions including closed-loop control, trending, and regulatory compliance are all supported by online instrumentation. Table B.12 summarizes the instrument uses by location.

MANAGEMENT

The operations staff members who work under the assistant superintendent are responsible for most of the instrument support. Operations work orders are computer-generated for calibration of a majority of the instruments. If the operators have difficulties performing the calibrations (e.g., the instrument will not calibrate), a work order is established for the instrument technician. Under some circumstances, the lab technicians may assist in troubleshooting instrumentation. The lab staff has historically been integral to instrument selection and testing. Figure B.3 depicts the city's structure as it relates to instrument support.

STUDY PARTICIPATION

CS-C evaluated four instrument types: ozone residual, pH, UV_{254}, and turbidity. This utility presented an opportunity to assess UV_{254}, an analysis that is rarely performed via online methods but which has significant potential future application because of rules governing organics removal. Table B.13 summarizes the instruments evaluated by location and frequency of bench-top verifications.

Table B.12
CS-C's instrument uses

Location	Instrument	Function
Raw water	Turbidity	Trending
Recarbonation	pH	Carbon dioxide feed control
Postozone	Ozone residual	Ozone feed control
Filter effluent	Turbidity	Trending
	Particle counting	Trending
Clearwell	pH	Sodium hydroxide feed control
	Turbidity	Compliance with 95th percentile
Backwash	Turbidity	Process
Finished water reservoir	Chlorine	Compliance
	Turbidity	Trending
Pilot-plant effluent	UV_{254}	Research

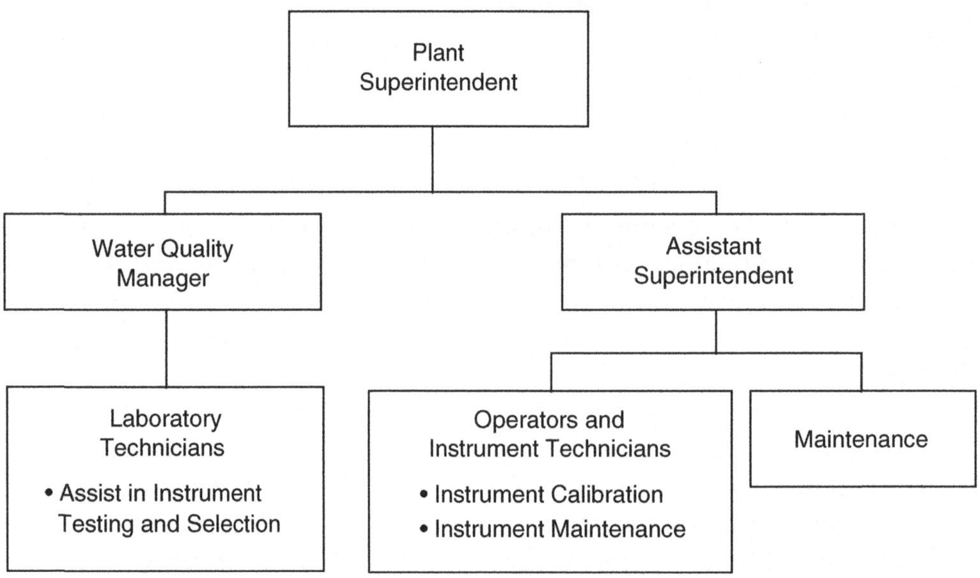

Figure B.3 CS-C's organizational structure as it relates to online instrument support

Table B.13
CS-C's participation

Instrument	Location	No.	Frequency of bench verification
Ozone residual	Cell 2	2	2 per week
	Cell 3	2	2 per week
pH	Recarbonation	2	2 per week
	Clearwell	2	2 per week
UV_{254}	Pilot plant	1	1 per week
Turbidity	Filter effluent	2	2 per week

CASE STUDY D

INTRODUCTION

Case study D (CS-D) is a 50-mgd conventional treatment plant treating a river source. As shown in Table B.14, the river source is fairly stable, exhibiting some fluctuations but not significant flashiness. Source water entering the plant first goes through a presedimentation basin. Following presedimentation, the flow goes to rapid mix where it is split into two trains, and the flow passes through the remaining conventional unit processes.

INSTRUMENTATION

Table B.15 identifies the location of online analytical instruments for this system, and Table B.16 displays the instrument support program for each online instrument.

Table B.14
CS-D's 1997–1998 raw water quality*

Parameter	Average	Minimum	Maximum
Turbidity (ntu)	7.7	2.2	15.5
Alkalinity (mg/L $CaCO_3$)	73	57	90
pH (SU)	8.0	7.6	9.2
Temperature (°C)	17.1	4	28
TOC (mg/L)	2.5	1.5	3.5

*Raw water quality is from the first 12 months of ICR data.

Table B.15
CS-D's online instrumentation

Location	Instrumentation
Raw	pH
Presedimentation	Turbidity
Rapid mix	pH, chlorine, SCD
Sedimentation	Turbidity (surface scatter), chlorine
Filtration	Turbidity, particle count

Table B.16
CS-D's instrument support

Instrument	Location	Support
Surface scatter	Sedimentation	Weekly: Clean chamber. Monthly: Calibrate with formazin. Semiannually: Replace tubing. Perform thorough cleaning with brush. For algae buildup, clean with 20% bleach solution as needed. Keep spare instrument on hand.
Turbidimeter	Filter effluent	Monthly: Perform cleaning and cube calibration checks. Quarterly: Calibrate with 40-ntu standard.
pH	Raw, rapid mix	Weekly: Clean.
Chlorine residual	Rapid mix, sedimentation	Weekly: Clean cells. Monthly: Change reagent. Semiannually: Replace all tubing.
Particle count	Filter effluent	Biweekly: Clean cells. Quarterly: Perform critical cleaning. Annually: Send in for calibration.

Instrument Challenges

Challenges at this water treatment plant include algae growth in the surface scatter chamber, which requires cleaning with a 20% bleach solution. Also, in the higher turbidity applications, floc can clog the effluent tubing and cause the instrument chamber to flood. This is handled by keeping the drain slightly open to keep heavy floc material moving. In CS-D's pilot facility, condensation has ruined the power sources and sensors on some turbidimeters.

Instrument Use

Table B.17 summarizes CS-D's instrument uses. The plant's lab group performs all compliance sampling; online instruments are generally used only for informational purposes and trending, or to set chemical feed rates.

STUDY PARTICIPATION

CS-D evaluated chlorine residual, turbidity, and pH as outlined in Table B.18. What makes this plant particularly interesting as a case study is that plant personnel spent a lot of time and effort developing standard operating procedures (SOPs) for their online combined filter effluent (CFE) turbidimeters. In fact, CS-D supplied its SOPs to regulatory developers, and portions were incorporated into a guidance manual. It will be interesting to see if their effort resulted in noticeable differences between this facility and others with less rigorous programs.

Table B.17
CS-D's instrument uses

Location	Online instrument	Function
Raw	pH	Chemical feed adjustment
Presedimentation	Turbidity	Trending
Rapid mix	pH	Trending
	Chlorine	Informational
	SCD	Chemical feed adjustment
Sedimentation	Turbidity (surface scatter)	Trending
	Chlorine	Trending
Filtration	Turbidity	Trending
	Particle count	Trending
Finished (combined filter effluent)	Turbidity	Informational purposes

Table B.18
CS-D's participation

Instrument	Location	No.	Frequency of bench verification
Chlorine residual	North postsedimentation	1	Daily
	South postsedimentation	1	Daily
Turbidity	North effluent	1	Daily
	South effluent	1	Daily
pH	North rapid mix	1	Daily
	South rapid mix	1	Daily

CASE STUDY E

INTRODUCTION

Case study E (CS-E) is an unfiltered surface water system. Three sources of water, one surface water and two groundwater, make up this city's available water resources. Under normal operation, water is collected from the river source as it emerges from a mountain range and is then conveyed many miles through pipeline before reaching the first customer. Sometimes the water upstream of the intake of the dam becomes excessively turbid as a result of rainfall, snowmelt, and related runoff events. Because the source is unfiltered, when these elevated turbidity events occur, the city either blends the river water or substitutes the source completely with groundwater. The blending takes place at the headworks. The city also has a well field, which has a capacity of 60 mgd for short periods of time and is used to supplement flow during peak demand (peak is approximately 90 mgd).

The river watershed is a protected source, which helps to maintain its pristine quality. The city owns some of the watershed property and has landowner agreements with 90% of the land it does not own, so source water protection activities can take place on most of the watershed property. Watershed inspectors patrol the watershed to identify and eliminate unsanitary conditions and to evict trespassers. Table B.19 shows the range of turbidity and temperature experience in 1997–1998.

INSTRUMENTATION

As an unfiltered surface water plant, more instrumentation is located in the distribution system than in other conventional plants. Table B.20 identifies the location and online analytical instruments for this system, and Table B.21 displays the instrument support program for each online instrument.

Table B.19
CS-E's 1997–1998 water quality*

Parameter	Average	Minimum	Maximum
Turbidity (ntu)	1.8	0.29	6.6
Alkalinity (mg/L $CaCO_3$)	16	14	20
pH (SU)	7.2	7.0	7.4
Temperature (°C)	11.0	6.7	22.0
TOC (mg/L)	1.0†	Below detection limit	1.6

*Water quality data obtained from the first 12 months of ICR data.
†Average TOC calculated from six above-detection-limit values.

Table B.20
CS-E's online instrumentation

Location	Instrumentation
Raw water	Turbidity
Distribution system	Chlorine residual, pH, fluoride

Instrument Challenges

During spring runoff and severe rainfall events, the river source turbidity increases rapidly. When this occurs, the source is blended with or completely replaced with groundwater. As a result of a severe turbidity event a few years ago, redundancy has been provided for raw and finished turbidimeters (raw is at the intake; finished is about a mile downstream at the headworks facility and follows source blending). In addition to turbidity events, seasonal algae blooms occur, impeding the instruments by plugging screens and coating probes and lenses.

Instrument Use

CS-E uses online instruments for multiple purposes—as an early warning system, for compliance with regulations, to make manual chemical feed adjustments, and for trending. Table B.22 summarizes the instrument uses.

MANAGEMENT

The water quality department at CS-E is responsible for all online instrumentation used to monitor water quality for the system. Instrumentation selection and purchase is the responsibility of the water quality coordinator. The water quality department has three mechanics, who are responsible for installing the instrumentation as well as performing equipment checks and calibration on a random basis. Although the mechanics are assigned stations, crossover between stations and mechanics is not uncommon. The city also has two water quality assistants, who perform routine instrument inspections and calibration at all locations (with the exception of the headworks). There are three operators for the headworks, and they perform all instrument support. The three headworks operators keep logbooks that detail calibration results and other instrument-related comments. Figure B.4 shows CS-E's organization as it relates to instrument support.

STUDY PARTICIPATION

CS-E evaluated fluoride and turbidity instruments. Because this facility relies on unfiltered surface water, turbidity especially is critical to its operation. Also, because the use of turbidity to determine when and how much blending of sources should occur is a unique application, this makes evaluating these instruments especially interesting. Table B.23 summarizes CS-E's participation, including the location of the instruments, the number of instruments of this type and at this location, and the frequency with which bench verifications were recorded for this project.

Table B.21
CS-E's instrument support

Instrument	Location	Support
Chlorine residual	Headworks, distribution system	Monthly: Check versus portable and span/reset. Check for leaks. Clean Y strainer. Check analyzer flow. Check carbon dioxide flow and storage (where relevant). Check versus chart recorder. Semiannually: Dismantle, clean, and service parts as necessary. Other: Zero with deionized water.
Fluoride	Headworks, distribution system	Weekly: Check versus portable; adjust as necessary if difference is >1 ppb; other assistant checks calibration 2 days later. (This means that this group performs two comparisons each week.) Monthly: Perform quality control check versus portable. Check for leaks. Clean Y strainer. Confirm flow rate. Check buffer level. Check versus chart recorder versus supervisory control and data acquisition (SCADA). Check electrolyte fill solution. Change formic acid, tubing, o-rings. Semiannually: Dismantle, clean, and replace parts as necessary.
pH	Headworks, distribution system	Monthly: Check versus portable and reset if required. Check for leaks. Clean Y strainer. Check flow. Check versus chart versus SCADA. Refill buffer. Bimonthly: Replace hose on finger pump. Semiannually: Dismantle, clean, and replace parts as necessary.
Turbidity	Headworks, distribution system	Headworks: Perform daily check versus bench; recalibrate using calibration cube weekly or as needed. Weekly: Recalibrate with calibration cube. Clean sensors with Simple Green.* Clean bubble chamber when algae blooms are bad. Backflush unit.

*Simple Green, Huntington Harbour, Calif.

Table B.22
CS-E's instrument use

Location	Instrument	Function
Raw water	Turbidity	Early warning signal for when blending with groundwater is required; trending
Distribution system	Chlorine	Compliance; manual chemical feed adjustments
	pH	Manual chemical feed adjustments; compliance
	Fluoride	Trending; compliance

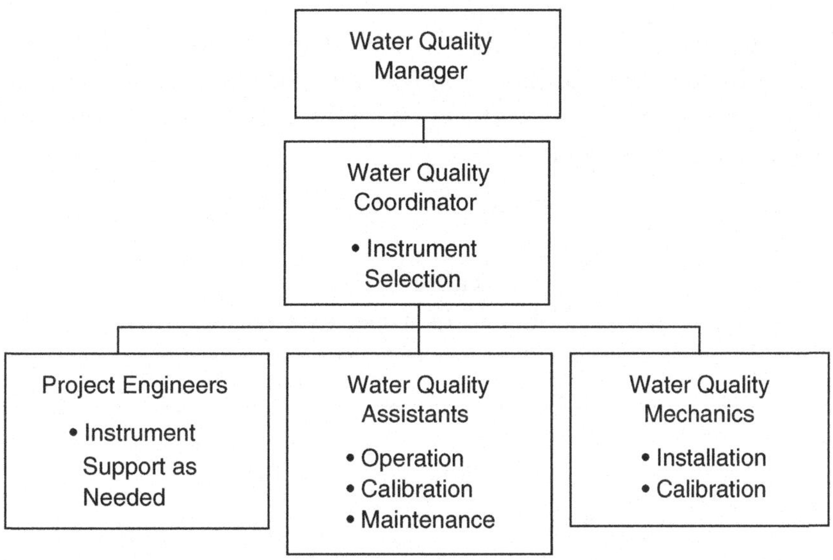

Figure B.4 CS-E's organizational structure as it relates to online instrument support

Table B.23
CS-E's participation

Instrument	Location	No.	Frequency of bench verification
Fluoride	Headworks	1	1 per week
	Distribution	2	1 per week
Turbidity	Raw	2	1 per day
	Finished	2	1 per day

CASE STUDY F

INTRODUCTION

Case study F (CS-F) is a conventional treatment facility with an annual average treatment of 41 mgd. Lake water is treated via conventional treatment through two parallel treatment trains. One of the trains has conventional sedimentation; the other has upflow clarification basins. Alum, chlorine, fluoride, and phosphate are the primary treatment chemicals used. Powdered activated carbon is used infrequently for taste and odor events.

The lake raw water quality is typically stable. Table B.24 shows the range of raw water quality recorded in 1998.

INSTRUMENTATION

Table B.25 summarizes all of CS-F's analytical online instrumentation by location. Typically, CS-F has adopted manufacturers' calibration procedures. Calibration and maintenance procedures do not vary as a result of location. Table B.26 summarizes online instrument calibration and maintenance procedures.

Table B.24
CS-F's 1998 water quality

Parameter	Average	Minimum	Maximum
Turbidity (ntu)	3.3	0.4	40
Alkalinity (mg/L)	120	111	140
pH (SU)	8.36	7.77	8.80
Temperature (°C)	11	0.6	27
TOC (mg/L)	2.0	1.8	2.4

Table B.25
CS-F's online instrumentation

Location	Online instrumentation
Raw water	Turbidity, temperature, alkalinity
Pretreatment	Chlorine residual, oxidation–reduction potential
Filter effluent	Turbidity (individual filters)
Combined filter effluent	Turbidity, chlorine residual, fluoride, and phosphate
Distribution system	Chlorine residual (two locations; CS-F wants to add turbidity measurements to distribution system)

Table B.26
CS-F's instrument support

Instrument	Location	Support
Phosphate	CFE	Weekly: Conduct bench-top comparison. Record and adjust reagent pressure. Record and adjust sample pressure. Check for leaking fittings, sample leaks, and reagent leaks. Monthly: Change reagents. Inspect and clean sample strainer. Quarterly: Clean colorimeter cell. Annually: Install annual maintenance kit. Replace colorimeter lamp. As needed: Replace reagent tubing module. Replace sample cell cover assembly. Keep a log of instrument maintenance.
Alkalinity	Raw	Weekly: Check titrant level (record amount added). Calibrate analyzer (input bench value). Check and record sample flow rate. Inlet valve setting. Monthly: Check operation of titrant pump and record flow rate. Set titrant pump seals. Quarterly: Replace titrant pump seals, drive belt, and o-rings. Wash and inspect pH electrode. Replace manifold and reaction cell o-rings. Clean sample cup. Replace pH electrode.
Turbidity	Raw, filter effluent, CFE	Weekly: Check lo out setting, should be 4 mA. Check hi out setting, should be 20 mA. Check outlet valve position, "degrees." Measure turbidity flow, should be 0.2–0.26 L/min. Monthly: Check with calibration cube. Clean sensor flow chamber. Lube o-ring. Quarterly: Recalibrate with calibration cube. Calibrate with formazin to 40 ntu. Recalibrate with calibration cube after 40-ntu calibration.

Treatment Challenges

Changes in this water treatment plant's source water quality can significantly challenge treatment. Online alkalinity instrumentation was installed to serve as an early warning of such changes in the lake, specifically as an indication for when alum needs to be increased. Bench versus online comparisons of the alkalinity instrument are performed weekly, and calibration is triggered when bench evaluations and online measurements differ by more than 3.2 ppm. This trigger was established by determining the in-house human variability inherent in performing online versus bench comparisons. The average variation was found to be 3.2 ppm, which is tolerable because it is likely a result of human error. Variability above this value could be attributed to calibration drift.

Frequent lake temperature changes are also challenging in water treatment at this plant. During the warmer months (June through September), the lake thermocline passes repeatedly back and forth over the intake. Online temperature helps to forewarn operations staff of temperature events that can upset the upflow clarification process. If these events are identified in advance, treatment can be adjusted to dampen the effect.

Instrument Use

CS-F does not use online instrumentation for closed-loop control of chemical feeds or other operational adjustments. In addition, most regulatory compliance results are measured via bench-top methods (i.e., fluoride, phosphate). Table B.27 summarizes instrument uses at this plant.

Table B.27
CS-F's instrument uses

Location	Online instrument	Function
Raw water	Turbidimeter	Trending
	Temperature	Trending; early warning associated with lake turnover and potential treatment challenges
	Alkalinity	Early warning associated with high chlorine demand
Pretreatment	Chlorine residual	Regulatory reporting of CT
Filter effluent	Turbidimeter	Trending
Combined filter effluent	Turbidimeter	Regulatory compliance
	Chlorine residual	Regulatory compliance
	Fluoride	Trending
	Phosphate	Trending, leak detection*
Distribution system	Chlorine residual	Trending

*The phosphate feed line is hidden and the online monitor is used to ensure that a leak does not go unnoticed.

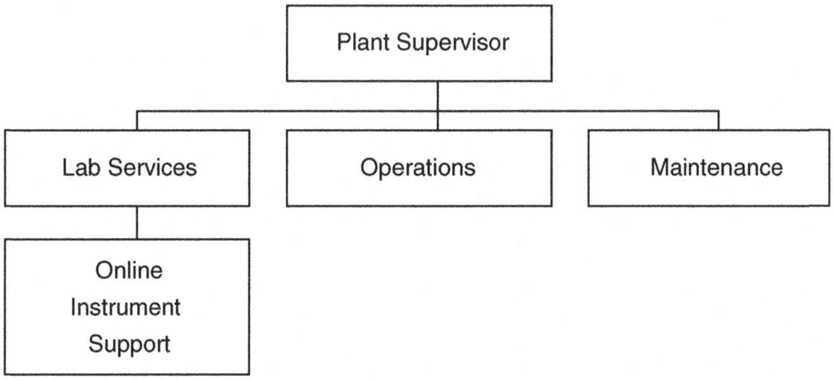

Figure B.5 CS-F's organizational structure as it relates to online instrument support

Table B.28
CS-F's participation

Instrument	Location	No.	Frequency of bench verification
Alkalinity	Source	1	Daily
Phosphate	Finished	1	Daily

MANAGEMENT

The laboratory staff is responsible for maintaining and calibrating all of the online instrumentation. If operations staff identifies a questionable reading or error signal, the lab is notified and responds to the event. Manufacturer calibration and maintenance schedules are typically adopted. Figure B.5 shows the organizational structure as it relates to online instrument support.

STUDY PARTICIPATION

CS-F sent historic data from its alkalinity instrument and collected phosphate data for this project. Bench comparisons are done daily on both the phosphate and alkalinity instruments and this information, along with maintenance information, was part of this study.

As mentioned previously, online alkalinity is measured to detect changes in raw water that require changes to alum feed rates. Phosphate is fed for corrosion control, and online phosphate (measuring orthophosphate) monitoring is primarily used to indicate that roughly the correct amount of phosphate is being delivered. Per the Lead and Copper Rule, plants must maintain a minimum total phosphate residual. A secondary purpose for these instruments, then, is to confirm that phosphate is being delivered. This is necessary because events in the past resulted in the appearance that phosphate was being fed, when in fact the feed system was leaking and the correct amount was not being delivered. All phosphate compliance samples are measured using bench-top methods. Table B.28 summarizes CS-F's participation in this study.

APPENDIX C
GUIDANCE FOR DEVELOPING A STANDARD OPERATING PROCEDURE

The example standard operating procedure (SOP) in this section can serve as a template that treatment plant staff can adapt to their own instrumentation. Comments on SOP development and areas where personalized information is needed are in all caps and enclosed in brackets to differentiate them from the text and sections that are part of the SOP.

TURBIDIMETER STANDARD OPERATING PROCEDURE

Manufacturer: ______________________	SOP Developed by: ______________________
Model: ______________________	SOP Revision/Date: ______________________
Instrument Location: ______________________	

INTRODUCTION TO TURBIDITY [SAMPLE INTRODUCTORY TEXT]

The cloudy appearance of water is caused by the presence of suspended and colloidal matter. In the water-works field, turbidity measurement (in nephelometric turbidity units or ntu) is used to indicate the clarity of water. Technically, turbidity is an optical property of the water based on the amount of light reflected by suspended particles. Turbidity cannot be directly equated to suspended solids because white particles reflect more light than dark particles, and many small particles will reflect more light than an equivalent large particle. As the primary tool currently used to assess a water treatment plant's ability to remove pathogenic organisms, turbidity is an important water quality parameter. In addition, turbidity measurements are commonly used at a number of locations within a plant to monitor treatment performance. For example, raw water quality can be assessed using a turbidimeter to determine coagulant doses, settled water turbidity can be used to evaluate coagulation/flocculation performance, and filtered water turbidity can indicate when a filter needs to be cleaned.

Table C.1 describes how turbidimeters are used at [YOUR PLANT NAME HERE].

ONLINE TURBIDIMETER OPERATION AT [YOUR PLANT NAME HERE]

The turbidimeter operational checklist (Table C.2 is an example) is located on a clipboard in the control room. Once a day [ENTER TIME, SHIFT, ETC.], complete this checklist for each turbidimeter. If problems are identified, please provide a description of how the issue was handled or notify [APPROPRIATE PERSON (I.E., STAFF TECHNICIAN)]. Note any issues identified in the logbook located [ENTER LOGBOOK LOCATION].

In addition to physical instrument checks, verify the online data regularly. The checks in Table C-3 have been established to confirm the validity of the data. These checks are programmed into the supervisory control and data acquisition (SCADA) system, and alarms have been established as shown in Table C.3.

Table C.1
Applications and associated locations for turbidity analysis

Application	Raw	Filter	Finish	Backwash
Automatic process control—alarm				
Automatic process control—chemical pacing				
Manual process control				×
Additional testing/monitoring				
Plant shutdown				
Compliance reporting			×	
Source water management				
Long-term trend analysis	×	×		

Table C.2
Checklist for routine inspection of online turbidimeters

Checklist item	Observed condition	Action required? (Y/N)	Followup action (describe)
Light-emitting diode (LED) readout on			
Readout variability			
Alarm/warning LEDs			
Valves open			
Flow (mL/min; measure weekly)			
Leaks			
Debris buildup			
Algae/microbial blooms			
Drains clear			
Other Items			
Initials/date			

Table C.3
Online operation checks

Parameter	Description	Alarm setting
Volatility	Running average of the variance of the online measure; N = 30 for measures every 15 minutes.	Volatility > ________
Interinstrument variation	Compare each finished water turbidimeter value against other finished water turbidimeters.	During normal operation, each instrument should not vary from the other finished water turbidimeters by more than ________.

CALIBRATION AND QUALITY ASSURANCE

Calibration, which helps to maintain the best instrument performance possible by routinely comparing the instrument reading versus a known standard solution, is a critical part of online turbidimeter operation.

If the instrument and known solution vary, calibration procedures will correct this variability.

Instrument performance goals will dictate the calibration frequency. For example, if an instrument performance goal of no greater than 0.10 ntu difference between online and bench measurements 99% of the time is established, calibration will occur at a much greater frequency than if expectations are of 0.20 ntu variability 95% of the time (as compared to bench measurements). Goal setting must also take manufacturer specifications, instrument application, and workload issues into consideration.

To establish calibration procedures, follow three steps:

1. Identify instrument operation goals (i.e., a bench versus online instrument delta of 0.2 ntu 95% of the time). Perform a check of the operational limitations specified by the manufacturer to ensure that goals are reasonable.
2. Collect and analyze data. Calibrate the online instrument and collect bench versus online data regularly to establish performance goals based on existing operations protocol. Analyze these data to identify the time span (i.e., calibration frequency) at which your operational goals are met.
3. Formalize and document protocol or establish new goals. If the calibration frequency established in step 2 is acceptable, develop a formal protocol. If it is unacceptable (e.g., it requires daily calibration), select a different goal with a frequency that meets your needs based on the data collected.

It is not necessary to repeat this data collection effort for each individual instrument; however, you will want to repeat it for instruments measuring different locations or different water quality, or both (see, for example, Table 5.1). Table C.4 is a sample summary table that describes different protocols.

Table C.4
Summary of data collection for calibration determination

Instrument location	Summary	Target accuracy
Raw	*Instrument application is trending.* Monthly bench comparison; calibrate when difference is ≥1.5 ntu or every 3 months, whichever occurs first.	<1.0 ntu difference 95% of the time.
Filter effluent or finished	*Instrument application is compliance.* Weekly bench comparison; calibrate when difference is ≥0.03 ntu or every 3 months, whichever occurs first.	<0.03 ntu difference 90% of the time.

DATA COLLECTION AND EVALUATION RESULTS

In this section, document the results from the data acquisition and analysis procedures used to estimate online instrument calibration frequency. Table C.5 is a template of a data table that may be useful (per Table 5.1 example).

Instrument Application ______________________________
Manufacturer's Specifications ______________ accuracy, ______________ precision
Operation Goals ______________________________

CALIBRATION PROCEDURE

Use the manufacturer's procedures as a starting point—these procedures typically cover the major points very well. In addition, explicitly delineate the following:

- Equipment necessary to perform the calibration procedure
- Required reagents
- Safety procedures for handling chemicals and glassware
- Specific protocol, outlined step by step
- Generation of the standard stock solution (in this case, formazin)
- Frequency or conditions that trigger a calibration event
- Management of calibration-related information

A complete SOP description of calibration procedures also requires a discussion of the "good laboratory practices" that provide confidence in the calibration protocol being repeatable by various staff members. Some sample good practices follow.

Table C.5
Raw water absolute value of the deltas as percentiles

Percentile	All data	Hours	Hours	Hours	Hours	Hours
0.05						
0.1						
0.2						
0.3						
0.4						
0.5						
0.6						
0.7						
0.8						
0.9						
0.95						
0.99						

- Use ultra-low-turbidity water to make turbidity standards. This water can be purchased or made by using a reverse osmosis water conditioning system for deionized and particle-free water.
- Make sure that all glassware used during calibration is very clean.
- Use an auto-pipette to measure formazin. These may come with a calibration kit or they can be purchased through a laboratory supplier.
- Use only freshly made formazin standards.

MAINTENANCE

Table C.6 lists specific maintenance requirements and the minimum performance frequency. Please perform the tasks at the frequency identified and record this information in the logbook.

If available and relevant, a maintenance software program or maintenance spreadsheet can be electronically linked to an Excel spreadsheet that lists maintenance requirements.

SPARE PARTS [OPTIONAL]

Generate and maintain a spare parts list and inventory regularly. Maintenance software can automatically perform inventories, but if such a program is not available, it is easy to set up a spreadsheet to track spare parts. Include the model number, description, and number of parts to keep in inventory in the spare parts spreadsheet. Most manufacturers' manuals contain a complete list of replacement parts.

Table C.6
Turbidimeter maintenance requirements summary

Procedure	Frequency
Clean lens	Monthly
Change lamp	Unscheduled—as needed
Change/clean tubing	Annually
Clean photocell window	Monthly
Replace formazin solution	Annually
Clean turbidimeter body	Semiannually
Others as identified	

SAFETY

The procedures preparation of the formazin calibration standard requires the following safety precautions:

- Wear gloves, eye protection, and lab coat or apron.
- Prepare the standard in a well-ventilated area.

The material safety data sheets included in appendix A are electronically linked to this document at [INSERT APPROPRIATE HYPERLINK HERE].

DATA MANAGEMENT

Utilities rarely document data management plans for online instrument support beyond the standard operating manuals for a utility's SCADA system. Develop protocols for standard data management practices. The protocols should begin at the point of an online measurement and end with its ultimate application and storage. The key features of such a plan include

- Data dictionary of online monitoring data parameters
- Data acquisition protocols
- Data quality assessment and flags
- Data storage and retrieval
- Standard reports
- Standard data exports and download procedures

REFERENCES

Burlingame, G., M. Pickel, and J. Roman. 1998. Practical Applications of Turbidity Monitoring. *Jour. AWWA*, 90(8):57–69.

Lewis, C., E. Hargesheimer, and C. Yentsch. 1992. Selecting Particle Counters for Process Monitoring. *Jour. AWWA*, 84(12):46–53.

Marshall, A., and A. Spiers. Online Corrosion Monitoring and Control in Continuous Food Sterilizers. *Jour. Mater. Perform.*, 31(11):47–51.

VanGelder, A., Z. Chowdhury, and D. Lawler. 1999. Conscientious Particle Counting. *Jour. AWWA*, 91(12):64–76.

ABBREVIATIONS

AwwaRF	Awwa Research Foundation
°C	degrees Celsius
$CaCO_3$	calcium carbonate
CFE	combined filter effluent
CPO	chief plant operator
CT	disinfection contact time
FE	filter effluent
GAC	granular activated carbon
ICR	Information Collection Rule
ISO	International Standards Organization
L/min	liters per minute
mA	milliampere
MCL	maximum contaminant level
mgd	million gallons per day
mg/L	milligrams per liter
N	number/count
NA	no frequency provided
ntu	nephelometric turbidity unit
pH	negative logarithm of the effective hydrogen concentration
PO_4	phosphate
ppm	parts per million
QA	quality assurance
QC	quality control
r^2	correlation coefficient
SCADA	supervisory control and data acquisition
SCD	streaming current detector
SOP	standard operating procedure
SU	standard unit
TOC	total organic carbon

UV_{254}	ultraviolet at a wavelength of 254 nanometers
Y/N	yes/no

IWA Publishing's authorised EU representative for General Product Safety Regulations is Diane D'Arras, 15 rue Duret, 75116 Paris, France, e-mail: safety@iwap.co.uk.

Printed and bound by CPI Group (UK) Ltd, Croydon, CR0 4YY
17/09/2026
02218799-0012